IN SEARCH OF AN INTEGRATIVE VISION FOR TECHNOLOGY
Interdisciplinary Studies in Information Systems

Contemporary Systems Thinking

Series Editor: Robert L. Flood
Monash University
Australia

A Continuation Order Plan is available for this series. A continuation order will bring delivery of each new volume immediately upon pulibcation. Volumes are billed only upon actual shipment. For further information please contact the publisher.

IN SEARCH OF AN INTEGRATIVE VISION FOR TECHNOLOGY
Interdisciplinary Studies in Information Systems

edited by

Sytse Strijbos
and
Andrew Basden

 Springer

Sytse Strijbos
Vrije Universiteit Amsterdam
The Netherlands

Andrew Basden
University of Salford
United Kingdom

Library of Congress Control Number: 2006920024

ISBN-10: 0-387-32150-0 (HB)   ISBN-10: 0-387-32162-4 (e-book)
ISBN-13: 978-0387-321509 (HB)   ISBN-13: 978-0387-321622 (e-book)

Printed in the United States of America.

9 8 7 6 5 4 3 2 1

springer.com

T
14
, I456
2006

CONTENTS

PREFACE

This book is the result of a quite exceptional research cooperation that has tried to chart a new course through the much-debated issues of Science, Technology and Society.

About the CPTS

In 1995 an international group of about fifteen scholars, junior and senior researchers from different disciplines came together in Amsterdam. This first meeting became the start of a formal cooperation between several universities and institutions in different countries. In 1996 the Philosophy Faculty of the Vrije Universiteit of Amsterdam (the Netherlands), the School for Business Administration and Social Science of the Technological University of Luleå (Sweden), and the Information Systems Institute of Salford University (UK) formed what is now called the Centre for Philosophy, Technology and Social systems (CPTS). At a later stage the School of Philosophy at the Potchefstroom Campus of the North West University (South Africa) and the Institute for Cultural Ethics (the Netherlands) joined the CPTS. These participants agreed to cooperate with the following objectives:

- ♦ to carry out an interdisciplinary research programme into the management and design of technology and social systems giving high priority to ethical and other normative issues;
- ♦ to promote the practical application of the research ideas developed at the Centre and vice versa to learn from the input from practice for further research work;
- ♦ to make available an international and interdisciplinary learning environment for doctoral students of the participating organisations.

Since its inception this Centre has initiated a number of activities. Its main activity has been its annual working conferences, at the beautiful venue of the Emmaus Priory, alongside the River Vecht in Maarssen, near Utrecht, Netherlands. At these week-long events, researchers present papers on their current research, receive comprehensive critical mentoring, and respond with ideas on how their research will be continued. This formula has proved very successful in generating a flow of high quality papers in international scientific journals, informing PhD research, and sharpening up ideas on a wide range of issues.

Charting the Course

The issues of Science, Technology and Society are both interdisciplinary and normative by nature. How should we tackle these issues? The CPTS did not start with a clearly developed conceptual framework enabling a fruitful interaction between the different disciplines involved - philosophy, systems thinking, and engineering - but it was anticipated that by working together such a framework would evolve through a process of learning-by-doing. This involves an ongoing reflection on the practice of our cooperation while thinking through such questions as: What is truly interdisciplinary research? What are the pitfalls that have to be avoided? How to manage the interaction between the different disciplines? Which disciplines do we need to involve? And, in general, what is the way to move ahead?

In searching for answers to these and similar questions, the participants in CPTS activities have often found common ground in the Dooyeweerdian school of philosophy, named after Herman Dooyeweerd (1895-1977), a Dutch philosopher in the Christian tradition linked to the Vrije Universiteit. We have often found it profitable to adopt, as a basis for an integrated approach, Dooyeweerd's ontology that there is a variety of aspects of any situation or activity (such as physical, biotic, analytic, lingual, social, economic, juridical, ethical and credal) that are irreducible to each other yet fundamentally intertwined. However, we have always welcomed and used ideas from many other thinkers - ranging from Stafford Beer and Peter Checkland in the systems arena, through to contemporary philosophers on technology and society like Jacques Ellul and Jürgen Habermas. The CPTS conferences have been marked by a healthy debate, an open mind for a variety of perspectives, and yet a thoroughly critical approach to each, including that of Dooyeweerd.

A main issue that has repeatedly engaged our interest and effort concerns that of normativity in relation to science and technology. Is there a firm basis to guide the development of science and technology for the benefit of 'the good society'? Searching for an answer, almost everyone connected with the CPTS accepted the Dooyeweerdian notion of intrinsic normativity as a starting point, which is to say that a normative framework for our activity in technology and management is not simply a matter of human construction, and thus is less susceptible to being dominated by vested interests, power relations or personal preferences.

For example, papers on agriculture, health, transport, educational, environmental sustainability and the like have been presented and discussed. It has become clear that the normativity in each area cannot be reduced to that in other areas.

The critical study of Science, Technology and Society has become a field in its own right and many are writing to say how technology is failing us. We have a whole spectrum, from studies of individual cases of failure right through to David Noble's (1999) view that technology as a whole has become a religion. Although such critiques on technology are useful, we believe that there is a strong need to turn our attention and critical effort to positive suggestions on how we might take our technological society forward. But these positive suggestions must not come from a swing back to technological optimism. They must arise from understanding - an understanding of the situation in which we find ourselves, an understanding of what technology could be, an understanding of our past that has brought us to where we are, an understanding of human beings, and an understanding of the responsibilities we hold (and to Whom we hold them).

This Book

During the first decade of the CPTS, papers on a wide range of topics related to information technology artifacts, their development, their use and evaluation, and IT in society, as well as other topics, have been presented and discussed. Since many of these contained high quality argument and innovative ideas, it was decided to write a book to bring them together and to present the work of the CPTS to the wider community. Doing this, it was hoped, might also enable us to reflect more clearly the nature of the common approach.

This book is the result of that process, and has the aim of providing an initial 'integrative vision for technology'. Thirteen of its sixteen chapters are versions of a number of these papers, sometimes updated with new material or to make more explicit how they contribute to the overall vision. The first chapter serves not only to introduce these papers, but it also sets out our current understanding of this vision, together with a conceptual framework that enables us to tackle the diversity of issues that we meet in the story that is of our technological civilization. Since the CPTS wishes its work to be open to critique from those holding other perspectives, the final two chapters are critical evaluations of the work from thinkers outside the CPTS.

The selection of papers for this book was made on the basis of academic quality (tested by peer review and publication in journals) and contribution to the overall integrative vision. How they relate to the overall vision and the precise reasons for the inclusion of each are explained in chapter 1. However, it should be noted that some of the papers, being the result of discussion at the CPTS from the mid 1990s, do not reference some of the latest literature; for example, the latest reference to Michael Jackson's work in critical systems is his 1991 book.

The reader might observe that many of the chapters make use of Dooyeweerdian philosophy, but do not do so in the same way. Each has found a different portion of Dooyeweerd's work useful, and each contains a stand-alone overview of the relevant portions, since they were published elsewhere, where Dooyeweerd's ideas were not known. As editors, we could have combined all these overviews into a single explanation of Dooyeweerd's ideas, but we decided against doing this on the grounds that each is tailored to the needs of its chapter, that doing so would lose the nuances of what each author felt to be important, and that it would also mask the variability in interpretation of his philosophy. One possible benefit of retaining several parallel explanations of Dooyeweerd's work, is that, should the reader find one difficult to follow, they might find another more readable.

Acknowledgements

At this place we express our thanks to several people who played a substantial role in the activities of CPTS. Donald and Veronica de Raadt, then of the Technological University of Luleå, Sweden, were leading figures in CPTS in the early days, followed by Dries de Wet and Anne-Marie Pothas, of Northwest University (formerly Potchefstroom University), South Africa, in the middle period. The late Arie Dirkzwager, Professor Emeritus of Educational Computing at Technological University of Twente, the Netherlands, though not professionally involved, attended every working conference in a personal capacity and provided immense encouragement and wise guidance.

The CPTS is grateful for the contribution made over several years by Henk Geertsema, holder of the Dooyeweerd Chair at the Vrije Universiteit of Amsterdam, who has given a number of introductory seminars on Dooyeweerd over the years.

The Editors.

THE CONTRIBUTORS

Andrew Basden is Professor of Human Factors and Philosophy in Information Systems at the University of Salford, U.K., with interests in knowledge based systems, user interface, virtual environments and multimedia. Previously he worked in the pharmaceutical, chemical, health and construction sectors, after obtaining his Ph.D. in computer aided design in 1975.
> *Contact: A.Basden@salford.ac.uk. Information Systems Institute, University of Salford, Salford, M5 4WT, U.K.*

Birgitta Bergvall-Kåreborn is Senior Lecturer at the Department of Business Administration and Social Science, Luleå University of Technology, Sweden, where she obtained her Ph.D. in a multi-modal approach to soft systems methodology in 2002. Her research interests include methods of organisational and information systems development, the interaction of information systems and organisations, assessment methods for information systems, and methods for user involvement.
> *Contact: Social Informatics/Division of Information Systems Sciences, Department of Business Administration and Social Sciences, Luleå University of Technology, SE-971 87 Luleå, Sweden*

Darek M. Eriksson is a management consultant in Sweden and senior lecturer at the Mid-Sweden University. He obtained his Ph.D. in Industrial Economics from Chalmers University of Technology in 2004. His research focuses on integrated and multifaceted modelling of management situations and Change Management in organisations. He is Member of the Board of the Swedish Operations Research Association, the managing editor for the International Journal Cybernetics & Human Knowledge, and on the board of other management journals.
> *Contact: darek_eriksson@hotmail.com. School of Informatics, Mid-Sweden University, Ostersund, Sweden.*

Gerald Midgley is a Science Leader at the Institute of Environmental Science and Research, New Zealand, where he leads an interdisciplinary research team, and is Adjunct Professor in the School of Management, Victoria University of Wellington. Previously he was Director of the Centre for Systems Studies, at the University of Hull, U.K. He is the author of *Systemic Intervention: Philosophy, Methodology, and Practice* (Kluwer/Plenum, 2000), editor of *Systems Thinking* (Sage, 2003) and co-editor of *Community Operational Research: OR and Systems Thinking for Community Development* (Kluwer/Plenum, 2004).
> *Contact: Gerald.Midgley@esr.cri.nz. Institute of Environmental Science and Research, 27 Creyke Road, PO Box 29-181, Christchurch, New Zealand.*

Anita Mirijamdotter is Senior lecturer in the Department of Business Administration and Social Science and Head of Research in Informatics and Systems Science, at Luleå University of Technology, Sweden, where she obtained her Ph.D. in multi-modal extension to soft systems methodology in 1998. Her research interests include participatory methodology for systems design in dynamic organisational settings, soft and normative aspects of information systems design and evaluation, the interaction of information systems for coordination and communication in organisational processes, and systems design as a learning process.

Contact: Anita.Mirijamdotter@ltu.se. Social Informatics/Division of Information Systems Sciences, Department of Business Administration and Social Sciences, Luleå University of Technology, SE-971 87 Luleå, Sweden

Carl Mitcham is Professor of Liberal Arts and International Studies at the Colorado School of Mines, a faculty affiliate of the Center for Science and Technology Policy Research at the University of Colorado, Boulder, and on the adjunct faculty of the European Graduate School in Saas Fee, Switzerland. Previously he held appointements at the Brooklyn Polytechnic University and Pennsylvania State University (where he served as Director of the Science, Technology, and Society Program). His publications include *Thinking through Technology: The Path between Engineering and Philosophy* (1994) and *The Encyclopedia of Science, Technology, and Ethics* (2005).

Contact: Stratton Hall 301, Colorado School of Mines, Golden, CO 80401, U.S.A.

Sytse Strijbos is the Chairperson of CPTS, and lectures in the Philosophy Faculty at the Vrije Universiteit, Amsterdam, with which he has been associated since 1977, and at the School for Philosophy at North West University, Potchefstroom Campus, South Africa, where he is a visiting professor. He is also Chairperson of the International Institute for Development and Ethics (Europe). His scientific career started at Philips Research Laboratories at Eindhoven where he worked for ten years in materials research and process technology. He has published widely on philosophical issues related to the systems sciences, and ethical problems of our technological world.

Contact: S.Strijbos@ph.vu.nl. Faculty of Philosophy, Vrije Universiteit, Amsterdam, 1007 MC, The Netherlands.

Johan van der Lei is Professor of Medical Informatics at the Erasmus Medical Center in the Netherlands. He is member of the editorial board of several medical informatics journals and chairman of the Dutch Medical Informatics Association. His research has concentrated on the development, evaluation, use, and impact of computer-based patient records with particular focus on primary care.

Contact: J.vanderlei@erasmusmc.nl. Faculty of Medicine and Health Sciences, Erasmus University, P.O. Box 1738, 3000 DR Rotterdam, The Netherlands.

Jan van der Stoep is Director of the Institute for Cultural Ethics in Amersfoort, the Netherlands, and post-doctoral researcher at the Vrije Universiteit in Amsterdam. His current interests lie in political philosophy and Science, Technology and Society studies and especially in issues of communication technology and multiculturalism. He studied biology at Wageningen University and philosophy at the Vrije Universiteit in Amsterdam, where, in 2005, he defended his Ph.D. thesis about Pierre Bourdieu and the political philosophy of multiculturalism.

> *Contact: j.vanderstoep@12move.nl. Instituut voor CultuurEthiek, Postbus 224, 6710 BE Ede, The Netherlands.*

Albert Vlug is Senior Advisor of the Netherlands' Health Government concering I.T. infrastructure. He studied computer science at the Technical University of Delft in the Netherlands and philosophy at the Free University of Amsterdam. He has developed proactive assessment methodology for information systems, in which both hard and soft aspects are interwoven, and has taken leading roles in the evaluation of major medical information systems, including the national drug safety system built at the department of Medical Informatics at Erasmus University, Rotterdam and the system developed under the European Commission's Information Society Technology Programme.

> *Contact: a.e.vlug@erasmusmc.nl. Department of Medical Informatics, Erasmus University, P.O. Box 1738, 3000 DR Rotterdam, The Netherlands.*

Mike Winfield is a Principal Lecturer in Computing and Information Systems at the University of Central England (UCE). He started teaching basic computer science and then moved into Artificial Intelligence where he developed an interest in Knowledge Elicitation. This interest was followed as a research topic and he received his Ph.D. from Salford University in 2000. This work provided an introduction to Dooyweerd's work and led to MAKE (see chapter 4) being developed.

> *Contact: Mike.winfield@uce.ac.uk. Business School, University of Central England, Perry Barr, Birmingham B42 2SU, U.K.*

SOURCES OF CHAPTERS

The editors and publisher wish to make the following acknowledgements and notes about the source of each chapter and when it was discussed at the CPTS Working Conferences. All diagrams in this book, though perhaps based on originals supplied by the authors, have been drawn manually by the editors in order to maintain uniform style.

Chapter 1 (S. Strijbos and A. Basden) 'Introduction: In Search of an Integrative Vision for Technology' is new material, written for this book.

Chapter 2 (A. Basden) 'Aspects of Knowledge Representation', is based on a version of Basden (1993), expanded to incorporate specific reference to Dooyeweerd, presented at CPTS 1998.

Chapter 3 (B. Bergvall-Kåreborn) 'Reflecting on the Use of the Concept of

Qualifying Function in Systems Design' summarizes three previously published papers:

Bergvall-Kåreborn B. (2001). The role of the Qualifying Function concept in systems design. *Systemic Practice and Action Research*, **14**:79-93.

Bergvall-Kåreborn B. (2002a). Qualifying Function in SSM modelling - a case study. *Systemic Practice and Action Research*, **15**:309-330.

Bergvall-Kåreborn B. (2002c). Enriching the model-building phase of Soft Systems Methodology. *Systems Research and Behavioral Science*, **19**:27-48.

Its content was presented at CPTS 2004.

Chapter 4 (M. Winfield and A. Basden) 'Elicitation of Highly Interdisciplinary Knowledge' is based on the material of Winfield, Basden and Cresswell (1995, 1996), with material from Winfield (2000) and some discussion of the knowledge management field. The core ideas were presented at CPTS 1995.

Chapter 5 (A. Mirijamdotter and B. Bergvall-Kåreborn) 'An Appreciative Critique and Refinement of Checkland's Soft Systems Methodology' is a combination of material from two published papers:

Bergvall-Kåreborn B. and Grahn A[1]. (1996). Multi-Modal thinking in Soft Systems Methodology's Rich Pictures. *World Futures*, **47**:79-92.

Bergvall-Kåreborn B. and Grahn A. (1996) Expanding the framework for monitor and control in Soft Systems Methodology, *Systems Practice*, **9**(5):469-495.

with some new material from more recent research. The contents of this chapter were presented at CPTS 2004 and its component ideas at several earlier CPTS conferences.

Chapter 6 (S. Strijbos) 'The Systems Character of Modern Technology' is partly based on 'Ethics for an Age of Social Transformation I: Framework for an Interpretation' by S. Strijbos *World Futures*, **46**:133-143 (1996), and is used with permission. World Futures' website is http://www.tandf.co.uk.

Chapter 7 (J. van der Stoep) 'Communication Without Bounds?' is based on a paper on 'hypermobility' presented at CPTS around 1995 and subsequently published as Van der Stoep and Kee (1997), but translating its ideas to the context of information and communication technology.

Chapter 8 (J. van der Stoep) 'Norms of Communication and the Rise of the Network Society' is an extended version of a paper presented at CPTS 1996 and published as Van der Stoep (1998).

Chapter 9 (A.E. Vlug and J. van der Lei) 'Evaluation of Systems in Human Practice' was presented at CPTS 1998, and then in its present form as 'Evaluation of a drug safety system' at the 42nd Annual Conference of the International Society for the Systems Sciences in Atlanta Georgia on July 19-24, 1998. The

---

[1] Grahn is Mirijamdotter's previous surname.

original version may be found in the Proceedings of the conference (ISBN 0-9664183-0-1). Copyright, the authors.

Chapter 10 (D.M. Eriksson) 'Multimodal Investigation of Technology-Aided Human Practice in Business Operations' is a reprint from an earlier published article, 'Multimodal investigation of a business process and information system re-design: a post-implementation case study' by D.M. Eriksson, *Systems Research and Behavioral Science*, **18**:181-196 (2001), copyright John Wiley and Sons Ltd., and is reproduced with permission. The ideas in this chapter were presented at CPTS 1998.

Chapter 11 (A. Basden) 'An Aspectual Understanding of the Human Use of Information Technology' combines and expands material from Basden (1994, 2001) and the core of its content was presented at CPTS 1996.

Chapter 12 (S. Strijbos) 'The Idea of a Systems Ethics' is a slighty revised version of 'Ethics for an Age of Social Transformation II: The Idea of a Systems Ethics' by S. Strijbos, *World Futures* **46**: 145-155 (1996) and is used with permission. World Futures' website is http://www.tandf.co.uk. Some of the content of this chapter was presented at CPTS 1998.

Chapter 13 (D.M. Eriksson) 'Normative Sources of Systems Thinking: An Inquiry into Religious Ground-Motives of Systems Thinking Paradigms' is an extended verison of an earlier published article, 'An identification of normative sources for systems thinking: an inquiry into religious ground-motives for systems thinking paradigms', by D.M. Eriksson *Systems Research & Behavioral Science*, **20**(6):475-487 (2003), copyright John Wiley and Sons Ltd., and is reproduced with permission. An early version of this chapter was presented at CPTS 1999.

Chapter 14 (S. Strijbos) 'Towards a 'Disclosive Systems Thinking'' is a slightly revised version of 'Systems methodologies for managing our technological society: towards a 'Disclosive Systems Thinking'' by S. Strijbos, *Journal of Applied Systems Studies* **1**(2):159-181 (2000) and is used with permission. This chapter was presented and refined at several CPTS conferences, including 1999.

Chapter 15 (G. Midgley) 'Reflections on the CPTS Model of Interdisciplinarity' is new material, written for this book, and was presented at CPTS 2004.

Chapter 16 (C. Mitcham) 'Technology and Systems - But what about the Humanities?' has been written specifically for this book, but the editors and contributor with to acknowledge that it draws on work contributed to the *Encyclopedia of Science, Technology, and Ethics* (New York: Macmillan Reference, 2005), and owes a debt to the research assistance of Adam Briggle.

Chapter 1

INTRODUCTION: IN SEARCH OF AN INTEGRATIVE VISION FOR TECHNOLOGY

Sytse Strijbos and Andrew Basden

1.   INTRODUCTION

Labels like 'interdisciplinary', 'multidisciplinary', 'transdisciplinary' are used for joint research efforts that aim for an integration of knowledge of different disciplines. What is exactly meant, however, usually remains obscure. Boden (1999) distinguishes six types of interdisciplinarity, which require different intellectual attitudes on the part of the researchers involved and different types of management of the research process. In order of increasing strength these six types of interdisciplinarity are:

1. *Encyclopaedic*: Several disciplines are made available but individual researchers are not forced to cross disciplinary boundaries (a university for example is usually structured as an encyclopaedic enterprise).

2. *Contextualizing*: One takes account of other disciplines, but without active research-cooperation with those disciplines.

3. *Shared*: Researchers share a common goal without a need for day-to-day cooperation, such as in the human genome project in which different research teams work together to achieve a complete mapping.

4. *Co-operative*: Researchers from different fields actively work together towards a common goal.

5. *Generalising*: A single theoretical perspective is applied to a broad variety of previously distinct disciplines (as in Wiener's cybernetics or Von Bertalanffy's General System Theory).

6. *Integrated*: Concepts and insights of each discipline contribute to the theoretical development of others.

Boden notes that the first, second and third type could be better termed multidisciplinary, rather than interdisciplinary. Only the sixth type, 'integrated interdisciplinarity', is genuine interdisciplinarity and is defined by Boden (1999:20) as "an enterprise in which some of the concepts and insights of one discipline contribute to the problems and theories of another - preferably in both directions. ('Both' or even 'all': there may be more than two disciplines involved.)" Boden (1999:22) suggests artificial intelligence (AI), a field in which she has been acknowledged as a leading scholar, exhibits integrated interdisciplinarity. Each of the streams in AI - symbolic AI, connectionism, and 'nouvelle AI' - has developed by borrowing concepts from other disciplines like philosophy, logic, psychology, and neurophysiology.

The essays in this collection are the result of dialogue and research cooperation between philosophers of technology, technologists (or engineers), and systems methodologists, which is described in the Preface. A need is felt in the diverse disciplines involved to broaden our understanding of technology, moving from focusing purely on technical issues to seeing technology in its human context. But the different contributors feel this need in different ways, which could result in fragmentation. So the contributors aim for an integrative vision for technology as an interdisciplinary framework that results in the different felt needs coming together.

This interdisciplinary cooperation seems to fit nicely with type six of integrated interdisciplinarity. Indeed, they might take Boden's ideas further, venturing into sociological and ethical areas as well as those cited by Boden, thus bridging the gap between natural science and humanities, and between theory and practice (Strijbos, 2004).

We first outline the nature of the need as felt by the three sets of contributors (section 2), then propose an integrative vision for technology (section 3), then, in section 4 we define four major interdisciplinary areas of research, within which the papers in this volume have been clustered as explained in section 5. Though the integrative vision on technology is of general interest, the work reported is centred on information technology, which is argued to be a good exemplar.

2.   THE NEED FOR AN INTEGRATIVE VISION FOR TECHNOLOGY

2.1   The need as felt by technologists

Technologists (i.e. those who engineer technologies and artifacts) often focus attention on the basic machine and its working principles. They abstract from the social-cultural and human context of technology, the web of human activities surrounding the machine, to solve technological problems. However, such a restricted approach of technology in engineering often fails, and technologists themselves are becoming more and more dissatisfied by the limitations of this approach.

A typical example borrowed from Pacey (2000:8) concerns apparently simple hand-pumps used at village wells in India:

"By 1975 there were some 150,000 of them, but surveys showed that at any one time as many as two-thirds had broken down. New pumps sometimes failed within three or four weeks of installation. Engineers identified several faults, both in the design of the pumps and in standards of manufacture. But although these defects were corrected, pumps continued to go wrong. Eventually it was realized that the breakdowns were not solely an engineering problem. They were also partly an administrative or management issue ... A breakthrough only came when all aspects of the administration, maintenance and technical design of the pump were thought out in relation to one another."

What in fact was needed was a broader view on technology, which took into account the social-technical context, the human use of the pumps and proper arrangements for their maintenance.

Examples of failure of technological projects can be found in many other domains of technology too. The disappointing returns on sizeable investments in information technology is well known, a dramatic example being the development of the London Stock Exchange's Taurus system. It cost over 75 million U.K. pounds took twelve years to develop, but was then abandoned (Green-Armytage, 1993). It has been argued that many such failures occur as a result of a narrow, engineering approach and the limitations in conventional design methodologies (Mingers, 1995).

2.2   The need as felt by systems methodologists

In a 1978 article about 'The Origins and Nature of 'Hard' Systems Thinking' the systems scientist Peter Checkland observed that variants of systems thinking that arose from technology such as 'systems engineering' and 'systems analysis' had not been very successful when applied to social problems. If one hopes for an improved approach, then it will be necessary, so he continued, "to understand the way in which the nature of engineers' thinking both shapes and limits systems analysis".

At issue here are what Checkland (1978:109) called the limits of the 'means-ends' scheme that is employed in technology and in 'hard' systems thinking. "The strand of systems thinking which derives from engineering ... [is] based upon the assumption that the problem task ... is to select an efficient means of achieving a known and defined end." Checkland (1970:14) agrees with the philosopher of technology Ellul that technology becomes a blind force if people think solely in terms of means and ends.

Then "nobody asks 'What's it all for?'; technique, transforming ends into means, exists for its own sake and technique will mark out the materially comfortable but inhuman state which is our future." Everything is subjected in that case to the ends that people have set with technology. The task Checkland (1970:17) gave himself, when he assumed his post at the University of Lancaster, was to break through the absolutization of 'hard' systems thinking and to develop a new approach in which room is given to human values: "our systems approach must encompass human values."

With his question about the nature and limitations of technical thinking and effort "to bring the human into Systems Engineering", Checkland (2000a:59) set the agenda for a discussion about the foundations of systems methodology that continued later, in the eighties and nineties, in the work of Jackson, Flood, Ulrich and others.

2.3   The need as felt by philosophers of technology

In an outstanding survey of the relatively young history of philosophy of technology. *Thinking through Technology: The Path between Engineering and Philosophy*, Mitcham (1994a) distinguishes an 'engineering philosophy of technology' from a 'humanities philosophy of technology'. The first starts with the practice of technology - the experience and problems familiar to technicians or engineer - and finds its way to philosophy. Here one can speak of an 'internal' approach, 'from the inside'. The 'humanities philosophy of technology', in contrast, starts with the human and social sciences, and approaches technology as a distinctive phenomenon of present-day culture. The point of departure is 'external', seeking answers to the questions with which technology confronts man and society in our time.

Upon surveying the history of the philosophy of technology, one finds that near the end of the nineteenth century, the emphasis was 'internal': important contributions were made by people engaged in the practice of technology. But as technology matured in a human context, the 'external' approach, by professionally schooled philosophers not familiar with technology like Ortega Y Gasset, Heidegger, Ellul, Mumford, became more prevalent.

However, the latter approach to technology may also be called 'external' in a second sense. Not only is technology approached from the outside but also this 'external' perspective seldom penetrates to the concrete core of technical practices. In general, philosophy of culture views technology as a 'black box' that remains unopened, so that one does not gain a view of how technology works or of its 'internal' structure. The 'external' perspective thus fails to understand the diversity and subtle coherence of technical practices and the issues with which technologists have to grapple, and cannot connect with much that engineer-philosophers think and do. As a consequence, it can provide little guidance to technologists and does not enjoy their respect. As Mitcham (1994a:63-64) points out, "Qua engineers, they [engineer-philosophers] do not question it [technical practice], and they commonly regard questions raised by others as distracting or beside the point. ... Humanities thinkers do not understand what they talk about, say engineer-philosophers." Between the two approaches to technology, the 'internal' and the 'external', a gap exists, and it is difficult to unify them.

3. OUTLINE OF AN INTEGRATIVE VISION FOR TECHNOLOGY

On what basis can the thinking of technologists, methodologists and philosophers be integrated? How is interdisciplinarity in this broad sense to be achieved? Two main streams of thinking have been employed in the work represented here in pursuit of an interdisciplinary framework. First, the concept of 'system' has been borrowed from the systems sciences enabling us to lay emphasis on the systems dimensions of modern technology. Second, fundamental notions have been used from the philosophy of Herman Dooyeweerd (1955) - such as meaning, modal aspect, qualifying and founding function, intrinsic normativity, disclosure - enabling us to develop a normative view on technology and society that accounts for the diversity of human life.

Based on these traditions of thought we developed a conceptual framework about the relationship between 'technology' and 'society', depicted schematically in Fig. 1. The figure shows five areas, representing different systems levels that can be distinguished in our integrative vision for technology. The two upper levels - directional perspectives and human practices - concern the human and social domain, while the three lower levels - socio-technical systems, technological artifacts, and basic technologies - represent technology. This order of levels signifies the founding role of technology for society and that the founding function of 'human practices' is the connecting link between 'technology' and 'society'.

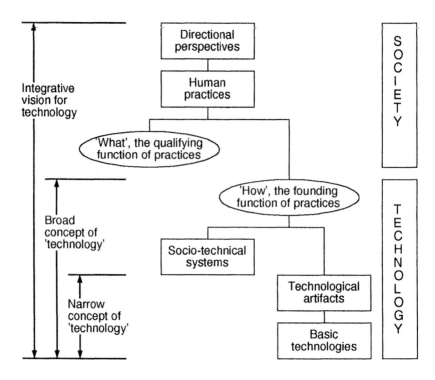

Figure 1. Schematic representation of technology as the founding of society

3.1   Society

It is typical for the integrative vision for technology that it does not focus on technology as such but situates technology in the normative context of human and societal aspects. In thinking through the normative aspects of technology, especially in the two upper levels in Fig. 1, one may distinguish between 'structure' and 'direction' (Wolters, 1986). 'Structure' refers here to the complex of normative aspects that obtains for a particular practice - including, for example, social, lingual, economic, aesthetic, juridical aspects. It is the notion of 'structure' that does justice to what is intrinsic. Complementary to 'structure', 'direction' indicates a spiritual perspective that guides the way in which people work out the 'structure' of a practice - spiritual motivation, ethical attitudes, ethos, worldviews and other things that deeply influence the more visible aspects of human practice. It can be either conscious or unconscious, and plays a guiding role in the life of individuals or communities and even influences the evolution of entire cultures.

In any practice, two of the aspects will stand out as particularly important. In the tradition of Dooyeweerdian thinking these are called the qualifying and founding aspects, both of which provide norms salient to the shaping of the practice. The qualifying aspect relates to the 'what' of a practice, making it different from other types of practice, in whatever context it may be found; e.g. medical practice differs from the commercial practice of an entrepreneur. The founding aspect relates to 'how' something is done. In contemporary, technology-rich society, this 'how' is expressed in the organizational framework and in the technical artifacts used. A classic example in which both can be discerned is the organization of work around steam power and the factory system, the two inventions that gave rise to the Industrial Revolution (cf. Pacey, 2000:18-23).

3.2   Technology

In defining 'technology' we found it useful to distinguish between a narrow and a broad concept. The narrow concept of technology regards technology as artifacts, which are constructed from basic technologies (the two lowest levels in Fig. 1), and which we use for a range of purposes. In ancient societies this was the only type of technology available, being largely constructed from natural materials and used in a natural setting, but in modern society another type of technology has emerged, for which we require a broader concept.

The broad concept of technology includes a third level, socio-technical systems, and recognises that the interaction between technology and society has fundamentally changed. Technology is more than the set of artifacts we find in our surroundings; it has fundamentally altered those very surroundings, and our daily lives have become dependent on it. This fundamental difference between traditional and modern technology is nicely illustrated by Pacey (2000:2) discussing the influence of the snowmobile on the lifestyle of the Eskimos. "With his traditional means of transport, the dog-team and sledge, [the Eskimo] could refuel as he went along by hunting for his dogs' food. With the snowmobile he must take an ample supply of fuel and spare parts; he must be skilled at doing his own repairs and even then he may take a few dogs with him for emergency use if the machine breaks down." Likewise, the modern motor car is of little use without a major infrastructure of roads, fuel stations, fuel refining, traffic control, traffic laws, policing, and the like.

The powerbase in our societies has shifted, moving away from the individual user of technology to those who are in control of the collective organizational frameworks in which we live.

3.3   Information technology as exemplar

Although the model above may be applied to any technology, the work reported here is restricted to information technology (I.T.). I.T. is not just an example of one technology among many, but an exemplar in which important issues that surround all technologies stand out clearly. I.T. demonstrates clearly how a given technology can have a wide range of application in different areas of life; for example, the word processor. It demonstrates that artifacts need not be physical in nature. It forces us to face the fact that design and development of the artifact must be accompanies by design and development of the surrounding human, organizational and other context (we refer to this as the information system (I.S.)). It challenges us to come to terms with the systemic character of much modern technology (e.g. the Internet) in which technology creates a new environment for human life. It also demonstrates clearly how technology can affect our directional perspectives, by raising our expectations and shaping our life habits; for example, multimedia presentations. In this way, a study of I.T. can throw light on a number of issues that remain optional or hidden in some other technologies.

More so than many other technologies, I.T. motivates us to work towards an integrative vision because it has become clear over the last thirty years that the levels are closely interwoven with each other and a narrow vision is detrimental to successful employment of the technology.

4.   INTERDISCIPLINARY AREAS OF INTEREST

It is the vision above, shown in Fig. 1, that has motivated and guided the authors of the chapters in this volume. Each level yields a different area of research and practice, and most areas are represented by three or four chapters. However, this does not imply that the work carried out thus far has addressed all levels to an equal degree; for the lowest level, basic technologies, only one chapter is presented. Since, however, even this chapter is orientated towards the creation of artifacts, we feel justified in including it in debate about the artifact itself. The chapters are thus clustered in four interdisciplinary areas of interest corresponding with the four levels that we have set out above: artifacts, socio-technical systems, human practices, and directional perspectives.

4.1   Area 1 - Artifacts and their Development

The first area of interest is concerned with the process of developing artifacts for human use, employing information technology, which might be of several generic types. For example, a medical database might have three components, its information storage constructed using the object-oriented data modelling technology, its user interface constructed in a windows style technology, and its remote data port constructed using HTTP technology (Hypertext Transfer Protocol: Berners-Lee, Fielding and Frystyk, 1996). Such generic types of I.T. must be developed, but our main interest lies in the creation of artifacts.

Since the artifact designed for human use comes into being by a deliberate process of development, it is no surprise to find that one of the most important issues in this area is that of the development process itself, which includes analysis of user situations, design, implementation, testing and the like. This has led to considerable resources being devoted to researching, generating and debating methodologies for development.

It has become clear over the years that the development process must concern itself with more than the technical artifact. The human and organizational context must also be developed, or the artifact will not provide the hoped-for benefits when in use. For example, along with the medical database, various types of documentation must be written and tested and training regimes must be designed and implemented. So this area of interest becomes quite complex, often involving the later areas of interest, especially those of usage and socio-technical systems.

In the early days, system development methodologies were modelled on those found in engineering, but over the last two decades I.S. methodologies have taken on a sophistication unrivalled in most other technologies, because of the need to be sensitive to the variety of 'stakeholders' that are affected by the arrival and operation of the information system. Soft systems methodology (Checkland, 1981) is one example of a well-developed answer to this need.

4.2   Area 2 - Socio-technical Systems

In addition to enabling the existence of such artifacts, however, I.T. itself has also become a socio-technical system. Artifacts and basic technologies (Area 1) can become, after a period of widening availability, part of a socio-technical infrastructure. For example, the Internet may be seen as an artifact which is used by individuals to obtain or send information, but is also an infrastructure that both enables and constrains the way many people live today. One challenge in this area of interest is to maximize the enabling and minimize inappropriate constraints. In order to grasp more fully the impact and nature of I.T., we have to embrace a 'broad' as well as a 'narrow' concept of I.T.

As with the motor car, with I.T. we find a similar diversity of elements of infrastructure, including not only the whole software and hardware industry, with its continued research and development, but also such things as protocols for transmission of information (from low level protocols like HTTP to ontological protocols like Knowledge Interchange Format (Genesereth and Fikes, 1992) and KQML (Finin, Labrou and Mayfield, 1997)), and legal structures related to the use of the Internet.

Thinkers like Ellul and Van Riessen have tried to throw light on technology from this viewpoint, and their work is developed below.

4.3   Area 3 - Human Practices

The third area refers to human practices that are enabled, hindered, or otherwise changed by use of the technological artifact in its systemic context. It is an area of concern because the mere arrival of I.T. does not automatically bring about improvements. Indeed, the failure rate of information technology is consistently estimated at over 50% (Gladden, 1982; Lyytinen and Hirschheim, 1987; Cotterill and Law, 1993; Butterfield and Pendegraft, 1996). Most of the failure is not technical in nature, but is often due to a variety of human and organizational aspects (Landauer, 1996), such as failure to deliver the system on time and to budget, or systems that, while they meet the agreed objectives, do not fulfil the users' real needs, or participants in development following their own hidden agendas to the detriment of the project as a whole.

It became recognized in the 1990s that research is needed in this area, so it is as yet a young field. Much that is written is an exposition of case studies, and there are few agreed frameworks for understanding the topic beyond variants of cost-benefit analysis on one hand and Actor-Network Theory (Latour, 1999) on the other.

Major challenge in this area are the unpredictability of benefits and impact of I.S. usage, especially over the long term, and how to reconcile benefits to one stakeholder with detrimental impact to another.

4.4   Area 4 - Directional Perspectives

Finally, the fourth area of interest is that of directional perspectives. These are worldviews taken by society about technology, and often take the form of cultural assumptions that we take for granted. Walsh and Middleton (1984) suggest we can approach these by asking four 'worldview' questions:

♦ "Who am I?" What is the nature, task and purpose of human beings in relation to this? For example, we might assume that we have a right to technological progress.

♦ "Where am I?" What is the nature of the universe we live in? For example, we might assume that life is inherently technical in nature.

♦ "What is wrong?" What is the basic problem or obstacle that keeps us from attaining fulfilment? For example, we might assume that being left behind in the technological race is a major problem.

♦ "What is the remedy?" How is it possible to overcome this hindrance to fulfilment? For example, as Pacey (2000:7) points out, we often assume the answer to be a technical fix.

Such worldview questions are directional because they act as "a guide to our life," (Wolters, 1993:4). Often they remain as unquestioned assumptions. The challenge in this area lies in recognising the perspectives that people hold, and understanding how they function directionally.

5.   OVERVIEW OF THE CHAPTERS

The book contains contributions relating to each of the four areas of interest and is thus in four parts. We could order these either from artifact to perspectives, or vice versa. In view of our call for a wider integrative vision, it would be interesting to start with directional perspectives and a vision of the 'good society', then consider human practices, then socio-technical systems, and conclude by discussing artifacts. However, we will follow the approach traditional among engineers, which is in the reverse direction, starting with the artifact and presuming that each level will bring benefit to the upper levels. In our integrative approach it must always be remembered that at each level the higher level should be taken into account; for example the developer of artifacts is assumed to already possess a view of good human practices, good socio-technical systems and good directional perspectives. So there is no need to first expose the view of a good society in explicit terms, but rather to foster an attitude among those working in the lower levels.

Part V of this book does not relate to a specific area of interest but contains critical reflections on the whole endeavour.

Part I: Artifacts and their Development

Chapter 2: A. Basden, Aspects of knowledge representation
Chapter 3: B. Bergvall-Kåreborn, Reflecting on the use of the concept of qualifying function in system design
Chapter 4: M. Winfield and A. Basden, Elicitation of highly interdisciplinary knowledge

Chapter 5: A. Mirijamdotter and B. Bergvall-Kåreborn, An appreciative critique and refinement of Checkland's Soft Systems Methodology

The first paper in this part is, strictly, within the area of interest of basic technologies, but has been included here because it anticipates the area of artifacts and their development. Traditionally, the main characteristics sought of knowledge representation technology have been sufficiency, efficiency and expressive power (Levesque and Brachman, 1985) but these focus the thinking on the technology itself and so do not contribute to an integrative vision for technology. Basden's contribution, 'Aspects of knowledge representation', suggests that if we are to adopt such a vision, then an additional characteristic, appropriateness, is also important. Appropriateness anticipates the aspects of both human life and socio-technical systems and seeks to embody them in the very kernel of the basic technologies with help from Dooyeweerd's philosophical understanding of what an aspect is.

In 'Reflecting on the use of the concept of qualifying function in systems design', Bergvall-Kåreborn discusses the notion of qualifying function in some depth and applies it, not to technologies, but to methodology for the modelling and design of artifacts and systems. She develops it into a precise and useful form and argues that the notion could be incorporated in Checkland's (1981) Soft Systems Methodology (SSM), to overcome two specific flaws.

Unlike Bergvall-Kåreborn's use of Dooyeweerd's thought to enrich an existing methodology, Winfield (2000) used it to devise a new methodology for knowledge analysis (MAKE: Multi-Aspectual Knowledge Elicitation). He was originally motivated by a concern that in building knowledge based systems it is difficult to ensure that all relevant knowledge had been elicited and encapsulated within the system. Since then MAKE has proved adept at widening the scope of analysis by stimulating participants to consider aspects they usually overlook, and a new look at MAKE is presented in chapter 4, 'Elicitation of highly interdisciplinary knowledge'.

Reflecting the structure of the book as a whole, which moves from lower towards higher levels, the chapter 5, 'An appreciative critique and refinement of Checkland's Soft Systems Methodology', is broader in scope than the other three. Checkland's SSM, and indeed his overall approach, have become widely known and used as an aid to analysing and designing human activity systems. But SSM is not without flaws and, in this chapter, Mirijamdotter and Bergvall-Kåreborn attempt to remedy some of these by employing Dooyeweerd's philosophical notion of normative aspects. Both this and chapter 3 are examples of how Dooyeweerdian concepts can be used to enrich methodology. This chapter makes a good lead into part II.

Part II: Socio-technical Systems

Chapter 6: S. Strijbos, The systems character of modern technology
Chapter 7: J. van der Stoep, Communication without bounds?
Chapter 8: J. van der Stoep, Norms of communication and the rise of the network
    society

The contributions in this section move from a 'narrow' concept of technology to a 'broad' one, not only taking into consideration the variety of technological artifacts, but also the socio-technical systems and networks by which they are interrelated. Unlike much work in this field, however, they try to develop this broad concept of technology in the light of an integrative vision for technology, in which also the relation between the techno-organisational foundation and the normative qualification of human practices is taken into consideration.

In chapter 6 Strijbos elaborates on the theories of Ellul and Van Riessen in order to discover the peculiar nature of modern technology and the impact of this technology on modern society. A description is given about how, by the use of the scientific-technological method, our human environment is turned into an all compassing socio-technical system. Strijbos also investigates the rise of information and communication technology and asks himself whether the advent of the computer might herald a fundamental break with the systems character of modern society.

In the next two chapters Van der Stoep elaborates further on the broad vision of technology and applies it to the world of human communication. In chapter 7 he argues that the design of the communication network engenders a longing for unlimited accessibility, which, he argues, is a myth. He explores how this myth unfolds itself in modern western society and how the drive to expand one's scope of action is supported and accelerated by the development of new technological means. The central focus of the paper is that when we are only concerned about the question 'how' to develop a communication network and not about 'why' we are doing it, the enormous increase in accessibility will be counterproductive, leading to hypercommunication instead of real communication.

In chapter 8 Van der Stoep is especially concerned with the way in which the communication network as a socio-technical system may be beneficially embedded in the human activity of communication. He describes how computer mediated communication changes the processes of writing and reading as well as the patterns of social interaction. But he also emphasizes that human communication, as social interaction by lingual means, has a specific normative structure that cannot be ignored, even while using the new interactive media of communication. In this way he not only tempers the high expectations and fears of postmodern theory, but also points toward a direction in which the communication network may be used and further developed in a fruitful way.

Part III: Human Practices

Chapter 9: A.E. Vlug and J. van der Lei, Evaluation of systems in human practice
Chapter 10: D.M. Eriksson, Multimodal investigation of technology-aided human practice in business operations
Chapter 11: A. Basden, An aspectual understanding of the human use of information technology

When information systems come into use, evaluation of the benefits (and conversely detrimental impacts) becomes essential. In 'Evaluation of human practices', Vlug and Van der Lei seek a framework within which evaluation methods may be developed. Using the national drug safety system in the Netherlands as an illustration, the authors discuss and compare the frameworks for evaluation offered by four systems approaches - 'hard', 'soft', 'critical' and 'multimodal' systems thinking - and conclude that the latter provides the most promising concepts.

Unexpected repercussions of I.S. use - many of which are adverse - are a major problem today. Eriksson, in 'Multimodal investigation of technology-aided human practice in business operations', seeks to throw light on this. The case investigated is a stock ordering and control system adopted by a vegetable wholesaler to reduce the number of sales personnel, and the attendant re-design of business processes that proved to be necessary. It is argued that not only the technology but also the human structures and relationships are to be 'designed'. The result of implementing the new system was unexpected: a reduction rather than an increase in profits. Eriksson discusses why this occurred and uses Multimodal Theory as a framework by which the adverse consequences can be accounted for and could have been predicted.

If the high failure rate of information systems in use is to be reduced, it is imperative that we gain an understanding of what it is to use technical artifacts in human practice. In 'An aspectual understanding of the human use of information technology', Basden develops a multimodal approach into a general framework by which to understand usage, derived partly from empirical studies and partly from philosophy. It provides a means of understanding benefits that recognises the diversity of human practices and indirect or long-term repercussions of use of information systems. Though placed in the area of interest that we have called Human Practice, the framework presented in this paper takes the socio-technical context into account.

Part IV: Directional Perspectives

Chapter 12: S. Strijbos, The idea of a systems ethics
Chapter 13: D.M. Eriksson, Normative sources of systems thinking: an inquiry into religious ground-motives of systems thinking paradigms
Chapter 14: S. Strijbos, Towards a 'Disclosive Systems Thinking'

Chapter 12 illustrates well how philosophical thinking can be applied to a state of affairs that we encounter today that has been forced upon us by the increasingly technological character of modern society. Referring to the systems character of technology discussed in Chapter 6, Strijbos proposes a new field of 'systems ethics'. This field is concerned not with the ethics of individual behaviour, but with the ethics of collective acting within complex technological systems. It is discussed that this kind of ethics requires a new approach, partly because human conduct interwoven with artifacts and embedded in organizational-technical systems is a new phenomenon of our age. Strijbos distinguishes different normative perspectives. In order to overcome the seemingly unresolvable tension between the opposite poles of an 'ethics of adaptation' and an 'ethics of liberation' he makes a plea for an 'ethics of disclosure'.

The final two chapters 13 and 14 show that systems methodologies are influenced by worldviews or normative, religious perspectives. Borrowing fundamental Dooyeweerdian ideas concerning religious ground motives that can be discerned in the history of Western thought, both Eriksson and Strijbos evaluate contemporary systems thinking. In chapter 13, Eriksson discusses how 'hard', 'soft', 'critical' and 'multimodal' systems thinking are not to be differentiated only on logical or historical grounds but that each is founded upon, and driven by, a different religious ground motive. In chapter 14, Strijbos develops a critical position with respect to 'hard', 'soft' and 'critical' systems thinking and, against this background, makes a proposal for what is called 'disclosive systems thinking'. He examines four normative principles that hold for it.

Part V: Critical Reflections

Chapter 15: G. Midgley, Reflections on the CPTS model of interdisciplinarity
Chapter 16: C. Mitcham, Technology and systems - but what about the
 humanities?

 While the authors of the chapters in Parts I-IV have worked together
on the 'integrative vision', the authors of Part V are from outside this
group, invited to make critical reflections on what has been achieved.
This was in order to foster debate with other academic communities and to
gain fresh input with suggestions on the way forward. For this purpose,
two leading scholars were invited to comment, from different fields,
systems thinking and the philosophy of technology.
 Gerald Midgley has gained a reputation in interdisciplinary thinking in
the systems sciences, and his current interests concern wider
environmental and societal issues and the place of technology therein. As
a 'critical friend', he directs comments in chapter 15 mainly the
interdisciplinary model outlined in this chapter rather than on the material
found in the individual chapters. He highlights five significant strengths,
yet he also points out two broad areas of concern: how much help the
model can give in ecosystems research and to what extent it can deal with
conflicts over normative beliefs and provide a basis for decisions to
abandon development of a technology.
 Carl Mitcham is a leading scholar in the philosophy of technology with
an encyclopedic view of its history and the ethical issues of technology
and society. He has shown sympathetic interest in this work for some
time and in chapter 16 provides a 'provocative epilogue'. Mitcham fears
that, despite our best intentions, if our argument starts from technology
and works upwards to directional perspectives (Fig. 1), it might still end
up treating technology as neutral and 'disembedded' from cultural issues.
 These two contributions have provided three key challenges for future
work. Midgley focuses on the model itself: we should explore how our
'integrative vision for technology' relates to ecological issues, and we
must find a principled basis for supporting or abandoning technologies.
Mitcham focuses on presuppositions underlying the model. Although
Mitcham is aware that the authors reject the neutrality of technology, he
challenges them to work out their stance more boldly. Instead of starting
with technology, they should start with directional perspectives and a
vision of the 'good society', consider human practices, then socio-
technical systems, and conclude with what this implies for technology.

Part I:
Artifacts and their Development

Chapter 2

ASPECTS OF KNOWLEDGE REPRESENTATION

Andrew Basden

1. INTRODUCTION

Programmers represent knowledge about the world in a computer language to produce an information system (IS) that may be used to help human beings in various tasks. For example, to make the process of writing contracts in the construction industry more open, the team led by Hibberd and Basden (1995) created the INCA knowledge based system (KBS) to write contracts that more accurately reflect the intentions of the parties than is possible by adapting standard forms of contract. On asking its user around three dozen questions and making inferences based on the answers obtained, the KBS selects appropriate clauses and combines them intelligently together to form the draft contract. An important feature of the KBS was that it encouraged the parties to explore the reasons for each clause, to question and change clauses. INCA 'contains' sophisticated knowledge about legal principles of contract and how to use it.

All this knowledge had to be represented in a programming language, or 'knowledge representation language', KRL. But what form of KRL is appropriate to such wide-ranging types of complex knowledge? Many of the issues in knowledge representation (KR) were explored in the 1980s but have now re-emerged in the field of knowledge management (KM), which seeks to represent wide-ranging knowledge relevant to an organisation, as a general aid to that organisation's prcesses.

There is one issue that had not been resolved by the end of the 1980s and still deserves our consideration. Reflecting on the experience of the 1980s, Brachman (1990) called for 'KR to the people'. KR had become a specialist field from which lay people were excluded. In KM, 'the people' are employees immersed in the knowledge that requires representation, and they should be the ones to represent it.

'KR to the people' is the issue we address in this chapter. It is part of a wider story, alongside the issues of knowledge elicitation and analysis that are discussed in chapters 3 to 5, and that of whether IS are effective and useful in aiding human tasks that is discussed in chapters 9 to 11. In this chapter, we are concerned solely with KR, and specifically with the basic facilities that enable 'the people' to represent knowledge easily.

1.1   Knowledge representation

In the process of representing knowledge, depicted in Fig. 1, a programmer (or 'knowledge engineer', KE) expresses knowledge in symbols offered by the chosen KRL (which can be textual or graphical). The knowledge so represented is that which is relevant to a domain of the world (such as contract authoring).

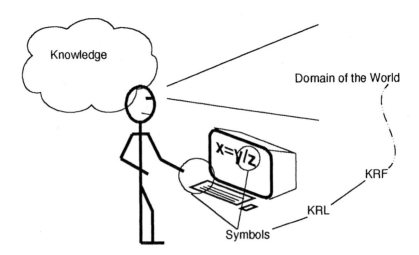

Figure 1. Knowledge representation as human activity

Traditional KRLs tend to be of certain basic types, which we will call knowledge representation formalisms (KRFs), developed out of pioneering work in the 1970s in artificial intelligence (AI) and computer science:

- ♦ production rules, in which knowledge is expressed as IF-THEN productions,
- ♦ logic programming (e.g. PROLOG), in which knowledge is expressed as predicate logic statements,

- ♦ functional programming (e.g. LISP), in which knowledge is expressed as nested function calls,
- ♦ frames and object oriented (OO) approaches, in which knowledge is expressed in terms of objects, classes, attributes and methods, and
- ♦ procedural languages like C, in which knowledge is expressed in terms of variables and instructions that manipulate them.

To design KRF software (for the KE to use) requires a number of decisions to be made about: types of 'thing' the KE is to be faced with (e.g. rules), ways these may relate (e.g. antecedent-consequent), types of property (e.g. truth values), types of activity (e.g. rule-firing), what valid inferences may be made (e.g. those of propositional logic), and types of constraint (e.g. that rules should not form an infinite loop). Such things are made available to the KE as atomic facilities but, at the bottom level, are implemented in machine code and memory structures.

But - as Brachman discussed - KR has proved more difficult than originally expected. Each KRF is appropriate for some types of knowledge and less appropriate for others; for example, it can be difficult to represent nuanced business strategy in any of the above KRLs. Moreover, if the domain involves several types of knowledge, the KRF used will be appropriate at some points, inappropriate at others. KR becomes esoteric expertise, complex and error-prone, and not suited to 'the people'. Expert KEs may be called in but they often fail to understanding the important nuances of the domain, so the ISs they build often fail to deliver the expected benefits when in use.

Thus the ideal KRF would be one that is appropriate to the whole range of types of knowledge that could conceivably be encountered.

This chapter, which is based on Basden (1993) but takes it further, discusses whether it might be possible to find such an ideal KRF. Using a concrete example of knowledge we might wish to represent, we briefly review the characteristics of various types of knowledge, which are discussed in Basden (1993). We outline the problems that arise in representing these using inappropriate KRFs and note that philosophy is needed to address them. After outlining some portions of a pluralistic philosophy, we show how they might be applied to develop more appropriate KRFs that enable diverse knowledge to be represented.

2. KNOWLEDGE REPRESENTATION PROBLEMS

The difficulties we are considering here are not those that arise from deficiencies in the KR software (which could be remedied by redesign) but those which seem inherent in the KR formalism. Consider the following scenario, illustrated in Fig. 2, in which we want an IS to help us keep track of birds' nests:

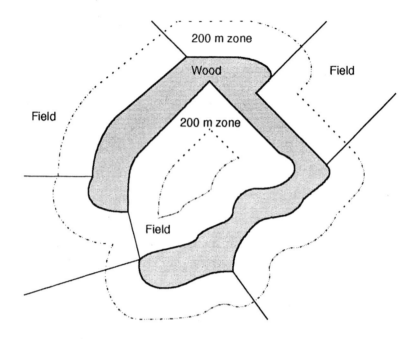

Figure 2. Nests, fields and woodland

♦ We have a terrain of rough pasture with fragments of woodland.
♦ The rough pasture is divided into fields.
♦ In the fields a species of bird nests.
♦ Eggs are laid in the nests, which (hopefully) will hatch into chicks.
♦ We want to keep track of, among other things, eggs in each nest.
♦ A species of fox lives in the fragments of woodland.
♦ At night the foxes emerge from the woodland to raid the nests (thus reducing the number of eggs in nests).
♦ The foxes tend to travel up to 200 m from the edge of woodland.
♦ We want to know which nests are in danger from foxes (so that we can, for example, watch and protect them).
♦ To do so requires obtaining permission from whoever owns the fields in which the nests are situated.
♦ So we need to know which fields each nest is in.
♦ But our knowledge of the dangers is limited (e.g. the behaviour patterns of the foxes) so we want also to search literature (on the Internet) to find relevant knowledge and link it to decisions we make.
♦ Finally, our usage of such a system might expand; for example, we might

need to start collecting and storing new information about nests that we do not currently anticipate.

All this knowledge must be represented in some KRF in order to build the IS we need. The reader will most likely agree that most of it can be understood intuitvely - the concepts of nests, fields, foxes, fragments of woodland, the concept of number of eggs per nest, that things change (eggs are laid and stolen), that certain nests are in certain fields which, conversely, contain those nests, that of ownership, the notion of fields and woodland as habitat, of the edge of woodland and of a zone up to 200 m from that edge, and so on.

Most knowledge that the KE must represent is explicit in the statement above, but some is tacit cultural knowledge. Our concern here, however, is not with whether knowledge is tacit or explicit, but with the diversity of concepts involved that seem intuitive or 'natural' to the lay person.

2.1 Representing knowledge to be used

Most ISs are designed to be used, and the best ones become part of the lifeworld of their users. After learning the features of the software itself, the user becomes familiar with the knowledge it holds, progressing from obvious, ordinary knowledge to the less obvious, exceptional knowledge, from seeing the knowledge as individual pieces to engaging with a nuanced, interwoven whole, from scepticism to trust.

What this implies is that not only should the knowledge of the domain encapsulated in the IS be accurate and understandable, but it should also be rich, nuanced and faithful to the recondite parts of the domain, so the user can rely on it without fearing lest some small but important piece of knowledge is missing. Much of the responsibility for ensuring this lies with the knowledge analysis and acquisition process (see chapters 3-5), but as Basden and Hibberd (1996) have argued, knowledge acquisition and representation cannot be separated, the latter often either hindering or stimulating the former. This means that any KRF the KE employs should facilitate faithful representation of a wide diversity of knowledge.

2.2 Appropriate KRFs

But which KRF should we use? As discussed more fully in Basden (1993), if we consider some of the 'aspects of knowledge' in our story above - items, relationships, values, changes, spatial and textual knowledge - we find that no KRF is appropriate to all of them.

An item (such as nest, field, fox) is a complex thing that persists over a period of time and has attributes that might change during this period. OO is well-developed for handling items of this kind, and PROLOG is very flexible in the way it breaks items down into statements about them. But procedural languages, in implementing items as structured collections of variables, are too rigid to allow new attributes to be added. Functional languages, though more flexible, hide the organic distinctness of items.

Relationships link items (such as "Field contains nest") and can be quite complex. They often have attributes (e.g. their start dates or strengths) and may even relate to each other (e.g. "The friendship between John and Jim was damaged by Jim's promotion"). Most real-life relationships have automatic inverses (e.g. if Field2 contains Nest5 then Nest5 is in Field2). Most are binary but not all are (an order line may be seen as a relationship between customer, supplier and product). Some types of relationships imply constraints, such as of hierarchy or prevention of loops ('directed acyclic graphs'). None of the KRFs above handles relationships well, most of them relying on implementing relationships as special attributes (known as pointers or keys) that hold identifiers to other items. This not only makes it tricky to implement genuine richness of relationships but it holds dangers (e.g. referential disintegrity, when the value in a key happens not to be a true identifier, or infinite loops when searching). As a result, for example, Blaha and Premerlani (1998:22) actively discourage consideration of inverses - but this inherently hinders the KE's ability to represent richness.

Values may be quantitative (such as 7, 'around 6') or qualitative (such as North, True, Loving), and are usually used as contents of attributes (e.g. 7 is NumberOfEggs attribute of a certain nest). Quantitative values may be incremented and imply notions like more, less and 'approximately near'. Procedural and OO KRFs offer a facility for simple values of both types, but logic programming has no inherent notion of quantity and must treat values as mere tokens, so it is cumbersome to increment or test for quantitative relations. (Practical logic programming KRFs handle quantity by means of hidden side-effects; see later.) The full richness of values extends beyond these simple notions to include, for example, degrees of precision (e.g. how near North?) and accuracy (how much to trust them). But few KRFs offer facilities to handle these.

Change is endemic in most applications - new nests being created, number of eggs in a nest changing, foxes raiding, and so on. Change can be either events or processes, and what triggers a change may be a particular time (e.g. spring time), or some sequence (egg laid then brooded then hatched), or some event (such as fox raiding) or on receipt of a message (such as alarm cry). Procedural and OO KRFs recognise the reality of change, with OO doing so in a slightly more sophisticated way, but logic programming has no inner notion of change and functional programming reduces change to the process of evaluating a function.

Spatial phenomena like shape, boundary, and a zone extending 200m out from the boundary, are important in many applications. Though the lay person finds such phenomena relatively simple to understand intuitively, none of the above KRFs offer spatial facilities. So none of them allows us to represent a spatially rich knowledge domain easily. (Software exists that encapsulates genuine spatial knowledge, including geographic information systems and Funts (1980) innovative 'geometric reasoner', but their facilities are not found in KRLs.)

Our final example is of information embodied in text or other symbolic form (diagrams). Our everyday experience of text - which is what we are concerned with in such an application - is of a complex structure that appears to be linear and hierarchical but is in fact interwoven in a more complex way (e.g. by both explicit cross references and subtle allusions), of interpreting and understanding it, of moving around it, of searching for approximately relevant material, etc. While most of the above KRFs offer simple text strings, none offer interwoven textual structure, approximate matching, aids to interpretation and the like that we encounter in everyday life. Software like word processors, Internet search engines and browsers offer such facilities, but none of this has found its way into KRFs yet.

2.3   Problems of inappropriateness

It is clear that none of the above KRFs make it easy to represent the everyday experience of all such aspects of knowledge, which is what may be necessary in a real-life application. Nor, indeed, does it seem that any KRF is able to approach the full richness of any single aspect. This leads to a number of problems in knowledge representation, many of them arising from pressure imposed by the restrictions in the KRFs.

a) **Richness** in knowledge may be reduced. For example, Blaha and Premerlani (1998) discourage using inverse relationships.

b) Knowledge can become **distorted** and so inaccurate or misleading. For example, if we can use only exact numbers, imprecision is masked.

c) Things for which the KRF is inappropriate, but which are acknowledged to be needed, tend to be implemented as **side-effects** - for example, PROLOG implements iterative change by means of the infamous 'cut-fail' technique. Side effects feel 'dirty', are difficult to learn and esoteric to use. 'The people' become confused and misled.

d) The knowledge base of the IS becomes more **complex**. For example, implementing inverse relationships by means of a pair of attributes (forward and back) introduces the need to keep them in synchronization even in the unforeseen circumstances. For example, the shape of the woodland in Fig. 2. and of the 200 m zone could each be represented as a list of (x,y) coordinates, which must then be managed. 'The people' should not have to bear such responsibility.

e) As a result, what seems 'natural' to our intuition requires extensive **KR effort** to represent. For example, to derive the buffered shape of woodland requires a complex trigonometrical algorithm that 'the people' are unlikely to know.

f) 'The people', under pressure to 'get something working', might inadvertently implement an **over-simplified** version of the aspect of knowledge, which later proves unable to support the full richness encountered in the domain. For example, a simple list of (x,y) coordinates cannot support a shape with holes in it (Fig. 2) - a very common mistake in spatial KR.

g) The **inner workings** of the implementation of an aspect become exposed and open to interference. For example, the 'key' by which a relationship is implemented might be given an illegal value.

h) Such interference can be **dangerous**. For example, an illegal value in a pointer might cause the whole computer to crash.

i) Finally, when a KRF covers several aspects of knowledge, the KE must decide on the **interface** between them.

Because of (a, b) the resultant IS can be inaccurate. Because of (b, c, e, f, g) KR takes on the nature of esoteric expertise. Because of (b, d, g, h) KR becomes more error-prone. Because of (c, d, e, i) the IS becomes more difficult to understand, which is problematic when, some years later, a new development team needs to upgrade it. Because of (a, b, f) some necessary upgrades are made impossible. Inappropriate KRFs, therefore, make Brachman's vision of 'KR to the people' difficult to achieve.

The conventionally recognised criteria for a good KRF, sufficiency, efficiency and expressive power (Minsky, 1981; Levesque and Brachman, 1985), do not address such problems. All they do is make (d, e) easier to achieve by expert programmers, and a focus on sufficiency and expressive power in particular often proves counter-productive because it suggests side-effects (c) are no problem.

2.4   Need for ontology

The issue of appropriateness presupposes a delineation of distinct aspects of knowledge and an understanding of the richness of those aspects. This in turn presupposes an ontological commitment to a vision of the nature of the diversity we encounter in the world. Tacit recognition of this may be seen in the proliferation, in the 1990s, of attempts to create 'ontologies' for KR and KM. But, with a few notable exceptions such as Wand and Weber (1995) who created a KR ontology based on Bunge (1977, 1979), most of these have only a tenuous link with philosophical ontology. Even Wand and Weber's ontology may be criticised because Bunge did not start from the notion of everyday lifeworld but from an assumption of an hierarchical system of levels informed mainly by the natural sciences (discussion of this must wait until another time).

We will explore an ontological approach to designing KRFs that derives from the philosophy of the Dutch thinker, the late Herman Dooyeweerd (1894-1977), which is based firmly in the everyday lifeworld.

3.   DOOYEWEERD'S ASPECTS AS BASIS FOR APPROPRIATE KNOWLEDGE REPRESENTATION

3.1   Modal aspects

We have spoken loosely about distinct 'aspects of knowledge'. Dooyeweerd also spoke of aspects, and indeed we shall find some correspondence between the two. On the first page of his *magnum opus*, *A New Critique of Theoretical Thought* (1955), he listed fifteen:

> "A[n] indissoluble inner coherence binds the numerical to the spatial aspect, the latter to the aspect of mathematical movement, the aspect of movement to that of physical energy, which itself is the necessary basis of the aspect of organic life. The aspect of organic life has an inner connection with that of psychical feeling, the latter refers in its logical anticipation (the feeling of logical correctness or incorrectness) to the analytical-logical aspect. This in turn is connected with the historical, the linguistic, the aspect of social intercourse, the economic, the aesthetic, the jural, the moral aspects and that of faith. In this inter-modal cosmic coherence no single aspect stands by itself; every-one refers within and beyond itself to all the others."

Unlike Bunge, however, Dooyeweerd related aspects to everyday experience. Immediately before the above text, we find:

"If I consider reality as it is given in the naïve pre-theoretical experience, and then confront it with a theoretical analysis, through which reality appears to split up into various modal aspects then the first thing that strikes me, is the original *indissoluble interrelation* among these aspects ..."

In a footnote he explained:

"[By aspects] are meant the fundamental universal modalities of temporal being which do not refer to the concrete 'what' of things or events, but are only the different modes of the universal 'how' ... For instance, the historical aspect of temporal reality is not at all identical with what actually happened in the past. Rather it is the particular mode of being which determines the historical view of the actual events in human society. These events have of course many more modal aspects than the historical."

Though Dooyeweerd began by setting out these aspects, he did not expect his readers to accept them uncritically but, at this stage, was merely setting the scene for a rigorous exploration not only of the structure of reality but also of the very nature of philosophical thought itself, by which we might address the structure of reality.

During his lengthy work, he uncovered the presuppositions underlying Western thinking since its root in Greek thinking 2,500 years ago, showed how they are responsible for the deep recurrent problems in Western thought, and then, starting from different presuppositions, developed a positive philosophy (in which aspects play an important part) to account for the nature of being, meaning, process, knowledge, normativity and time, which is breathtaking in its scope and novelty. This provides an underpinning for the aspects he introduced on his first page.

Many of the chapters of this book discuss parts of Dooyeweerd's philosophy; so here we will outline only those parts that help us understand appropriateness of KR and aspects of knowledge.

3.2   Roles of aspects

Many thinkers make reference to aspects, maybe under a different name, largely as irreducible categories that should be separately taken into account in analysis (such as Checkland's E's that are discussed in chapter 5). It was one of Dooyeweerd's insights however (Henderson, 1994:37-8) that aspects are not just categories of meaning but, being grounded in a transcendental notion of Origin, possess a modal character which enables creation to Be and Occur. Each aspect provides or enables:
- ♦ distinct ways in which things may be meaningful
- ♦ distinct ways in which things make sense (rationality)
- ♦ distinct modes of being

- ♦ distinct ways of functioning
- ♦ basic kinds of properties
- ♦ ways in which things can relate
- ♦ a distinct sphere of normativity.

Each aspect is a whole constellation of meaning, which accounts for the richness of meaning we experience. But at the centre of each is a kernel meaning can be grasped with the intuition, but never fully grasped by theoretical thought. In approximate terms the kernel meanings of the aspects (reversed in sequence, and with some names different from above partly because Dooyeweerd varied the names he used):

- ♦ Pistic aspect: faith, commitment and vision
- ♦ Ethical aspect: self-giving love, generosity, care
- ♦ Juridical aspect: 'what is due', rights, responsibilities
- ♦ Aesthetic aspect: harmony, surprise and fun
- ♦ Economic aspect: frugality, skilled use of limited resources
- ♦ Social aspect: social interaction, relationships and institutions
- ♦ Lingual aspect: symbolic signification
- ♦ Formative aspect: formative power, achievement, history, technology
- ♦ Analytical aspect: distinction, conceptualizing and inferring
- ♦ Sensitive aspect: sense, feeling and emotion
- ♦ Biotic (organic) aspect: life functions, integrity of organism
- ♦ Physical aspect: energy and mass
- ♦ Kinematic aspect: flowing movement
- ♦ Spatial aspect: continuous extension
- ♦ Quantitative aspect: discrete amount.

As an example, the constellation of the lingual aspect includes (and is expanded later):

- ♦ ways in which things may be meaningful: e.g. writing, rather than marks on paper (which would be sensitive) or poetry (which would be aesthetic), medium rather than substrate (which would be physical)
- ♦ ways in which things make sense: e.g. flow of argument or narrative
- ♦ modes of being: e.g. word, sentence, paragraph, diagram, noun, verb, bullet list, abbreviation, hyperlink
- ♦ ways of functioning: e.g. speak, write, read, translate, search.
- ♦ basic kind of properties: e.g. emphasis, clarity of wording, connotation
- ♦ ways in which things can relate: e.g. cross-reference, synonym
- ♦ sphere of normativity: e.g. words should be drawn from a vocabulary, text should obey laws of syntax.

3.3   Characteristics of aspects

Aspects are irreducible to each other, in that the meaning of one aspect can never be derived from, nor explained solely in terms of, that of others. Attempts to do so denature it and end in antinomy or paradox (Dooyeweerd used the example of Zeno's paradox to illustrate reduction of the kinematic to the spatial aspect). What this means is that each aspect is equally important and that the beings, functions, etc. that different aspects make possible must be considered separately. By contrast, systems theory sees aspects (levels) as 'emergent' from others by virtue of structure and behaviour rather than meaning.

Though irreducible to each other, there is also an "indissoluble inner coherence" between aspects (Dooyeweerd, 1955,I:3), which is of two types, analogy and dependency. Analogy means that each aspect contains echoes of the others (e.g. logical entailment (analytic aspect) echoes causality (physical aspect)). But it is inter-aspect dependency that more directly concerns us here. Aspects depend on each other for their full meaning. Dependency defines the order in which the aspects are normally listed, with pistic being 'latest' and quantitative, 'earliest' (note, not 'highest, lowest', to avoid any connotation that any aspect is more important than others). It is in two directions: an aspect depends on earlier ones for its implementation and on later ones for its full meaning. Thus, for example, the lingual aspect could not operate without formative structure, and symbolic signification would be of limited use were it not to allow us to communicate and thus relate socially. Note that dependency does not entail reducibility.

The earlier aspects are determinative, in that the outcome of response to their laws (e.g. laws of physics) is (largely) determined, whereas the later aspects are normative (e.g. laws of linguistics), allowing and enabling freedom.

Finally, no aspect is absolute, in the sense of being a self-sufficient foundation of meaning for the others. Nor is the entire suite of aspects taken together absolute, but all depend ultimately for their meaning on a Divine Origin.

One important corollary of this is that, as Dooyeweerd explicitly stated in (1955,II:556), "the system of the law-spheres [aspects] designed by us can never lay claim to material completion. A more penetrating examination may at any time bring new modal aspects of reality to the light not yet perceived before. And the discovery of new law-spheres will always require a revision and further development of our modal analyses." This is because to make a distinction among aspects we have to function in the analytic aspect, and if this is non-absolute then it is impossible to rely on the distinctions that it produces, such as a suite of aspects. However, this is not a recipe for universal skepticism since, as Dooyeweerd claimed, the aspects may be known by the intuition. Intuition, to Dooyeweerd, is not some mysterious nor instinctual way of knowing, but an awareness in which all aspects play their part.

This relates to Dooyeweerd's theory of everyday life (lifeworld), in which everything involves (i.e. can have meaning in) every aspect. For example, as I write this, I am functioning primarily in the lingual aspect, but I am also forming my thoughts (formative), distinguishing what to write (analytic), seeing the screen (sensitive), breathing (biotic), and exerting force on the keyboard (physical). I am also trying to give the reader what is due (juridical), to present a coherent argument (aesthetic), within a limited number of pages (economic), and I am treating the reader not as mere information sink but as fellow human being (social aspect).

4. A DOOYEWEERDIAN APPROACH TO APPROPRIATENESS

In like manner, use of an IS is multi-aspectual lifeworld (see chapter 11). Therefore, in principle, every aspect is important, and knowledge from every aspect should be represented in an IS. If this is to be done easily by 'the people', then the KRF they use should facilitate, and not hinder, the representation of knowledge of each aspect. We will now discuss how we may apply these portions of Dooyeweerd's philosophy to understand appropriateness, and thence to propose an approach to designing KRFs. First, we employ our knowledge of some aspects to design aspectually-oriented KRFs, then consider new aspects, and finally discuss the integration of such KRFs into a single, overall KRF.

4.1 Aspects of knowledge and individual KRFs

What we called aspects of knowledge correspond quite closely with Dooyeweerd's modal aspects (the use of the word 'aspect' in both cases is fortuitous since at the time the author knew nothing of Dooyeweerd):
- items: analytic aspect (distinct concepts)
- relationships: formative aspect (formed structure)
- values: quantitative aspect (discrete amount)
- spatial: spatial aspect (continuous extension)
- text: lingual aspect (symbolic signification).

We will examine these five aspects and see how, by reference to some of their philosophical roles, we can design a KRF for the corresponding aspect of knowledge. Referring back to the design decisions in section 1:
- Mode of being indicates types of 'things' to make provision for.

- Way of functioning suggests activity to provide as procedures.
- Basic types of property suggests attribute types.
- Way of relating suggests type of relationship.
- Aspectual rationality indicates inferences to be built into our KRF.
- Normativity suggests constraints that would be meaningful.

We work these out tentatively (and without any attempt at completeness) for several aspects as follows:

For the quantitative aspect (discrete amount),
- 'Things': integers, ratios, fractions, proportions, etc.; also types that anticipate later aspects such as 'real numbers' for the spatial aspect.
- Inferences: those of arithmetic
- Attributes: quantity, accuracy, units, etc.
- Relationships: greater and less than, sets, etc.
- Procedures: incrementing, scaaing, statistical functions, etc.
- Constraints: e.g. a given quantity remains that quantity until changed
- Example KRF: APL. Example software: calculators, statistical packages.

For the spatial aspect (continuous extension),
- 'Things': shapes, lines (straight or curved), areas, regions, dimensional axes, etc.
- Inferences: those of geometry
- Attributes: size, orientation, distance, side (in, out, left, right), etc.
- Relationships: spatial alignments and arrangements, touchings, crossings, surroundings, toplogy, etc.
- Procedures: join, split, stretch, deform, rotate, overlap, expand, etc.
- Constraints: e.g. boundaries should not have gaps
- Example KRF: Funt's (1980) geometric reasoner. Example software: geographic information systems, computer aided design, drawing packages.

For the analytic aspect (distinction),
- 'Things': concepts, objects, labels to identify things, etc.
- Inferences: those of formal logic
- Attributes: truth values, difference and sameness, etc.
- Relationships: contradiction, logical entailment, identity, etc.
- Procedures: e.g. distinguish, deduce
- Constraints: e.g. principle of non-contradiction, entity integrity (as in relational databases)
- Example KRF: PROLOG. Example software: cognitive mapper.

For the formative aspect (formative power),
- 'Things': structuring, relationships, plans, means and ends, goals, intentions, power, etc.
- Inferences: those of synthesis
- Attributes: feasibility, efficacy, version, strength (as of a relationship), etc.

- Relationships: means and ends, the purpose of something, sequence of operations (history), part-whole, etc.
- Procedures: form, compose, relate, revise, undo (to previous version), seek, effect a meaningful change (change a state), plan, etc.
- Constraints: e.g. referential integrity
- Example KRF: production rules. Example software: planning software, project control software.

For the lingual aspect (focusing on syntax and semantics, of text r.t. graphics):

- 'Things': nouns, verbs, etc.; words, sentences, etc.; bullet lists, headings, cross references, quotations, etc.; word roots, languages
- Inferences: those of syntax, semantics, etc.
- Attributes: tense, case, emphasis, cultural connotation, etc.
- Relationships: synonyms, antonyms, opposites, cross references, rhymes, thesaurus relationships, etc.
- Procedures: write, draw, understand, send message, text search, find equivalent meaning, translate, etc.
- Constraints: e.g. words to be within a vocabulary (spell checkers), syntax to conform to a grammar (grammar checkers)
- Example KRF: SGML, HTML. Example software: word processor, knowledge representation software.

Even though the features we have mentioned for each aspect are just examples, they exceed those supported by most extant KRFs. Yet they are not esoteric, but are features encountered in everyday living, of which some are already found in practical software oriented to the lifeworld). So why are they not implemented in KRFs? Such an aspectual approach to the design of KRFs might yield more comprehensive KRFs, which would go a long way to fulfilling the aim of 'KR to the people'.

But a number of issues still require our attention.

4.2   Change and qualitative values

Basden (1993) suggested that change (events, processes) is an aspect of knowledge, but there is no Dooyeweerdian aspect of change. Instead, to Dooyeweerd, change is bound up with time and inherent in every aspect.

The aspect of knowledge, values, originally included qualitative values (e.g. truth values), but has been aligned solely with the Dooyeweerdian quantitative aspect above. Like change, qualitative values are different for each aspect, and are constituted of its basic types of property. Table 2 gives a few examples of types of change and of qualitative for several aspects.

Table 2. Examples of change and qualitative value for certain aspects

Aspect	Example Change	Example Qualitative Value
kinematic	movement and flow	(mostly quantitative)
physical	causal reactivity	state (gas, liquid, solid)
biotic	growth and decay	alive v. dead
sensitive	stimulation and response	colour
analytic	from premise to conclusion	true v. false
formative	creation, construction and shaping	difficulty
lingual	utterance and understanding	which language
... up to ...		
pistic	the act of committing	sacred v. profane.

4.3   New aspects?

Basden (1993) discussed only four aspects of knowledge, conflating items and relationships into a single aspect and ignoring text. The question was raised whether there were other aspects of knowledge, and even whether those aspects are valid. At the time he had no other basis for answering this than intuition and observation of practical difficulties of trying to reduce one to another.

Dooyeweerd's aspects provides a sounder basis on which to answer these questions. Not only do the characteristics and philosophical roles of aspects provide a systematic way of considering aspects of knowledge and their corresponding KRFs, but it also provides, on one hand, a basis for splitting aspects of knowledge should that be necessary, and, on the other, a way of identifying possible new aspects and their KRFs, such as the economic or juridical. Some of the work on KM 'ontologies' may be seen as reaching for KRFs of later aspects. For some, such as the economic aspect, a useful initial KRF might be constructed, but for others, such as the ethical and pistic, it is not yet clear what shape a KRF would take.

4.4   Integrating KRFs

So far we have linked KRFs separately to different aspects. But any given application will involve representation of knowledge of several (if not all) aspects, and so the KRF we seek does not centre on a single aspect, but includes KRFs for all relevant aspects integrated together. To achieve this we might combine several aspectual KRFs, but we do not want to end up with a confusing muddle. Dooyeweerd's theory of inter-aspect relationships, however, can help us: just as the aspects relate to each other by dependency and analogy, and yet remain distinct, so aspectual KRFs may be integrated into a system by dependency and analogy and yet remain distinct. That dependency does not entail reducibility is an important philosophical insight by Dooyeweerd, because it provides us with the possibility of integrating aspectual KRFs.

Inter-aspect dependency in the foundational direction implies that a KRF for any aspect will require the facilities offered by a KRF of an earlier aspect. For example, a KRF for the lingual aspect will require at least the (formative) ability to relate and structure things like sentences, and to make distinctions such as for purpose of emphasis (analytic aspect). This suggests that the architecture of an aspectual KRF will involve a module for each aspect, and that if aspect Y is later than aspect X, then a KRF which has a module for Y will also need a module for X.

Dependency in the anticipatory direction suggests that some of the things, functions, constraints, etc. in aspect X are only meaningful in anticipating a later aspect, Y. (For example, irrational numbers are meaningful by reference to the spatial aspect, and linguistic pragmatics by reference to the social.) Since some of the later aspects have yet to be opened fully, it is likely that there are things in earlier aspects whose full significance is yet to be recognised. This means that when we implement a KRF module for any aspect (possibly excepting the pistic), we should make it possible to extend its things, procedures, constraints, inferences, etc. by such means as call-back hooks. (This is, of course, standard practice in reusable software today, but an awareness of aspects might be a useful guide to this practice.)

By its nature, inter-aspect analogy is more difficult to clearly define. It may be that extensibility features similar to those required for anticipatory dependency will be adequate, but since their exact form is likely to be different, it would be wise to keep them separate.

4.5 Discussion

These are only initial suggestions that illustrate how KRFs might be designed by considering Dooyeweerdian aspects and their constellations of meaning. We have considered only a few aspects, and the approach still has to be properly explored, tested and refined, but there is reason to believe that, if reasonably full account is taken of the constellation of meaning of each aspect, it will address the problems we identified in section 2.3 that result from inappropriate KRFs:

a) **Richness** need no longer be reduced inadvertently because all that is meaningful in an aspect is provided.

b) For a similar reason, **distortion** need no longer occur.

c) Implementation by **side-effects** is no longer necessary if all relevant aspects are included.

d) KB **complexity** is reduced to that which is intrinsic to the domain and not artificially increased by extraneous programming needed to implement other aspects.

e) KR becomes more 'natural' because all that is intuitive in each aspect is provided by atomic features and the KE no longer has to implement them, so less **KR effort** is required.

f) Thus there is less danger that a **simplified version** will be implemented.

g) Since all that is meaningful in each aspect is provided as atomic features, the **inner workings** of their implementation are no longer exposed.

h) So the attendant **dangers** are reduced.

i) In an integration of aspectual KRFs **interfacing** between the KRFs is already built-in and the KE no longer has to take responsibility for it.

The vision of an integrated, multi-aspectual KRF presented here is a long-term one. Not only are individual extant KRFs impoverished versions of the full aspectual KRFs envisaged, but the challenge of integration has yet to be met. It may be possible, during a period of research and experience, to gradually build up a sophisticated KRF that is appropriate to many aspects. The author's Istar KBS software (Basden and Brown, 1996) started to be developed along these lines.

However, there are several problems with this proposal that need to be addressed over the longer term. First, it is not always clear what shape a KRF might take for the later aspects, such as the juridical. Before we can tell whether the proposal made here has any merit, we should attempt to define KRFs for every aspect. There is evidence that this may be possible and even desirable, because features from later aspects - such as the juridical feature of copyright notices and authentication checks - are becoming important in practical software even today. At present, they are being introduced on an ad-hoc basis, which might insert problems for later; it may be that Dooyeweerd's aspects can enable a more systematic and future-proof approach.

Second, if we are to integrate aspectual KRFs, meaning in one aspect has to be translated to that of another. How this may best be done is not yet clear. The given suite of aspects proposed by Dooyeweerd does at least motivate us to do this, and his notion of the two types of relationship between aspects (dependency and analogy) can inform such consideration.

Third, if a complete set of facilities were implemented for every aspectual KRF would not the resulting software package be completely unwieldy? Though this has not been researched yet, there are two reasons for believing this might not be so. One is that the aspects are readily learned, and intuitively grasped; as Winfield (2000) and Lombardi (2001) have found, it becomes second-nature to consider them in any situation; see chapter 4. The other is that since the aspects have irreducible meaning, then it is possible that the symbols that express that meaning can be designed to be relatively independent of each other.

5. CONCLUSION

This is the only chapter in this book devoted to information technology as such. It has been concerned with the basic knowledge representation formalisms from which information systems may be constructed. It has been argued that, if we are to achieve Brachman's (1990) call for 'KR to the people' - a call that has yet to be fulfilled - then the KRF should be appropriate to all the diversity of knowledge found to be relevant in the domain of interest. This means that the facilities available to the knowledge engineer should enable all the richness of the domain knowledge to be fully represented, without distortion, without undue complexity and yet without simplification, without making the whole process more prone to errors than it need be, and without the need for esoteric knowledge of how to represent one aspect of knowledge in terms of another. We have argued that one way to achieve this is to ensure that the KRF gives every aspect of knowledge its due.

Aspects of knowledge, it has been further suggested, may be derived from the modal aspects proposed as a philosophical ontology by Dooyeweerd (1955), and that from these appropriate KRFs may be designed. All human life, including using an IS to aid our tasks, is multi-aspectual, involving every aspect. Appropriateness of a KRF is constituted in its ability to enable the knowledge engineer to give aspects of the everyday life of using an IS their due.

Modal aspects are irreducible spheres of meaning that are more than categories; they enable being, doing, relating, properties, norms, and each of these indicates a different portion of a KRF. Thus our proposal is that we should first devise a KRF for each aspect, and then integrate them together. There are philosophical reasons to believe (in the general case) not only that every aspectual KRF is needed (because of aspectual irreducibility) but also that integration is possible (because of an "inner indissoluble coherence" between the aspects).

A number of limitations in the proposal have been discussed. These may be taken to motivate future work - the development of KRFs for every aspect, of interfaces between aspects, and of ways to make the whole easy to use.

There is a surprising twist in this proposal. There is a general assumption that the inner design of technology is a purely 'technical' matter, divorced from the 'human' side, and that for a 'socio-technical' approach the technical must somehow be plugged into the human or social. The assumption may even be detected in chapter 1 of this book, in that the 'integrated vision for technology' sees the human, social context as outside the technical. But here we have seen that the very inner design of information technology - that central part which enables us to write programs - is intimately connected with all aspects of human everyday living. The human, social context is, or should be, already inside the very guts of information technology.

Chapter 3

REFLECTING ON THE USE OF THE CONCEPT OF QUALIFYING FUNCTION IN SYSTEM DESIGN

Birgitta Bergvall-Kåreborn

1. INTRODUCTION

Knowledge creation and knowledge sharing between traditional organisational boundaries are becoming more and more common in our contemporary world as new organisational forms, such as "network organisations", "virtual organisations", and "meta-organisation" (Groth, 1999), develop with increasing frequency. The importance of crossing boundaries successfully highlights the importance of recognizing and negotiating different perspectives or discourses at individual and group levels, which systems scholars such as Churchman (1971, 1984), Ackoff (1979b), and Checkland (1981) brought forth already in the 1960s and 1970s. These differences must be taken seriously, if these new organisational forms are to function well. The aim of this chapter is to continue the work of these system scholars, especially that of Checkland, by presenting an intellectual tool that facilitates the process of unfolding underlying rationalities related to problem definitions and improvement suggestions in system design situations.

System design is a broad concept that represents the systemic process by which design visions are confronted with physical and mental constraints and by that moulded and modified until a change that transforms an existing situation into a future one can be achieved. The idea of betterment or progress is almost inherent in the concept of system design since the future state, at least at the outset, is seen as an improvement on the existing situation. System design methodologies differ, however, in their recommendations of how betterment and progress is defined and achieved and for whom it is intended.

The existence of different perspectives and discourses, due to characteristics such as culture, religion and gender, is largely something to be celebrated. It is something from which we (stakeholders or participants) can learn. But before we can share and learn from each other we first need to communicate our perspectives. Consequently, there is a need for conceptual tools that can help us to express and explicate this diversity. One such tool can be found in the framework of Herman Dooyeweerd (1997) which presents fifteen dimensions covering many of the basic and traditional sciences and ranging from arithmetical and physical to social, ethical, and pistic. Related to this framework is a concept called the 'qualifying function' which aims to identify the leading or qualifying function of a certain system or activity.

The structure of the paper is as follows. First, the concept of qualifying function, and its related framework, is introduced and three different application areas are explored for the concept as a tool for system design. Then Soft Systems Methodology (Checkland, 1981, Checkland and Scholes, 1990a, 1999) is presented as a context for the concept of qualifying function. Following this, the tool is applied in an empirical study with the purpose of exploring its practical applicability as well as its strengths and weaknesses. Finally, some conclusions and implications for further research are provided.

2.   EXPLORING THE QUALIFYING FUNCTION AS AN INTELLECTUAL TOOL

The qualifying function is defined as the dimension that qualifies or characterizes a system, or an activity, (Dooyeweerd, 1997) in that it influences the character or form of that system. In a discussion about how it is possible to justify why a social system pursues one objective over another, Vickers (1959) says that, when faced with such questions, the system can only answer, or stop a regression of similar questions by answering, "Because that is the sort of system I am" or by saying "Because that is the value I place upon it," which, according to Vickers (p.100), is only another way of saying the same thing. I agree with Vickers in that the norms of a social system, to a large extent, determine the character of that system. And it is my argument that the qualifying function chosen for a social system reflects these norms as well as influences them. Further, since design focuses on the new and not yet created, on reality as it can become, or be made to be, this implies that in system design the task is to decide which function(s) best relates to the design vision of the designer(s). It also means that, even in the analysis of an existing situation, it is important to be able to view this situation from many different perspectives since perceptions on reality affect design visions for the future.

Further, the concept of qualifying function is related to a framework consisting of fifteen modalities or aspects related to human experience and thought as well as to many of the traditional sciences, in that the qualifying function of a particular system is represented by one particular modality. Starting from the more determinative and ending with the more normative, the modalites are as follows: the arithmetic, spatial, kinematic, physical, biotic, sensitive (psychic), logical, historical (cultural), lingual, social, economic, aesthetic, juridical (political), ethical and pistic. For further definition of these dimensions, see Fig. 1. For an illustration of the modalities and their meaning in context, see chapter 5.

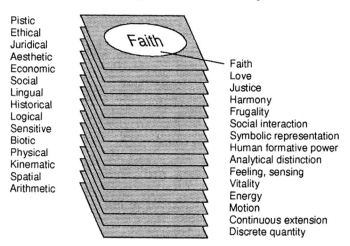

Figure 1. The modalities and their nuclei

The aim of the modal framework is not to exhaustively describe all of reality, but rather to indicate the rich variety in human life and remind us that this richness cannot be reduced into one or a few single dimensions. When studying a situation or entity, the framework can support both an analytical and a systemic approach. Following an analytical approach, the aspects can be studied in isolation from each other in order to clarify the specificity of each aspect. Following a more systemic approach, the relationship between the modalities comes into focus - i.e., their inseparable and mutual coherence, as well as their interrelatedness.

In order to clarify the meaning of each aspect, the elements are linked to a nucleus, as depicted in Fig. 1. Hence, the lingual modality is characterised by its kernel symbolic representation, while the social aspect is characterised by social interaction.

In the following, I will explore three application areas where the concept of qualifying function can be used as a tool for system design. The first application area for the concept is its use as an inspiration or stimulus to help people in a particular problem situation to view that situation from new and different perspectives and through that identify and remove self-imposed constraints.

The importance of being able to perceive situations from different perspectives has been argued by many authors (Ackoff, 1993; Checkland, 1981; Flood and Jackson, 1991a) and cannot be overemphasized in a design situation. It also constitutes an important foundation for identifying and removing self-imposed constraints, which according to Ackoff (1993), provide two out of three steps in creative behaviour. The third is an exploration of the consequences of having done so. This, however, often requires that we can change, break away from, or at least question our present frame of mind or worldview, something that has proven to be a very difficult task. As an example of how hard it can be to question previously unquestioned worldviews, consider the well known example of the shift from the Aristotelian to the Copernican worldview related to the structure of the Universe.

Earlier and somewhat similar examples of frameworks or tools aimed at broadening our perspectives are Morgan's (1986) eight metaphors of an organization, and Total Systems Intervention, where Flood and Jackson (1991a) relate different systemic metaphors (similar to those of Morgan's) to various systems methodologies. They argue that these metaphors can highlight aspects of a particular problem situation and hence guide the choice of appropriate systemic problem-solving methods. Or as Midgley (1997) says, "Participants can use them to think in different ways about the issue with which they are concerned." In addition, Flood and Jackson argue that consciously looking at problem situations through different metaphors should help a manager think creatively.

To illustrate how the concept of qualifying function can help us tease out different perceptions of a certain system and break away from ingrained notions and frames of mind, I shall use the example of a hospital as an intellectual experiment. If I choose the pistic modality as the qualifying function of a hospital, I picture how this modality could be thought of as governing the activity of healing. An example of this is how sickness was perceived in prehistoric times. At that time sickness was considered to be a punishment from the gods or the result of evil persons' witchcraft, and the symptoms were usually seen as evidence that the body had been taken over by an evil spirit. Hence, the remedy was sorcery and magic. This view of sickness and health, I would argue, is closely related to the pistic modality.

In Sweden, on the contrary, hospitals have, by tradition, been viewed as governmental organizations with missions focusing on biological curing and healing. It is the biological modality that, to the largest extent, has guided these organisations in both their practical and theoretical endeavours. In recent times, however, other modalities, such as the ethical, social, and aesthetic, have gained recognition as important for the healing process. It could therefore now be argued that the Swedish focus should shift from the biological and that the leading function for hospitals should be to give love and care rather than to save lives. However, considering the large cutbacks in hospital budgets, necessitating policies limiting the elderly to a once a week shower and fresh air, it could also be argued that it is the economic modality, rather than the biological, ethical or aesthetic, that increasingly influences hospitals. Cut backs have forced some hospitals to shut down wards, refuse admission to people in need of their services, and issue public statements saying that they can no longer guarantee the safety of their patients. Even if these three perceptions are viewed as radical interpretations, they still illustrate how the modalities help us to view a situation or a system from different perspectives.

It could, of course, also be argued that the concept of qualifying function, while being based on the fifteen modalities, can create a boundary for our thoughts and hence constrain rather then free our minds. While I have to agree that this is a possibility, my defence is that all concepts, tools, methods, or methodologies have this inherent weakness. While they help us and guide us in how to think, what to do, and how to do it, they also possess the possibility to constrain us. As soon as you say "Do this", it can also be interpreted as saying "Do not do that". However, I argue that the assistance offered by the modal framework far exceeds any constraint they might introduce. Further, this potential in weakness can be limited by awareness of its existence and incorporation of many diverse methods into any design process.

The second application area of the concept of qualifying function is its usefulness in assessing different design alternatives and their consequences. This can be achieved by choosing different qualifying functions for the system to be designed and then analyzing how these choices would affect the characteristics and structure of the system. If we take a hospital as an illustration once again, limiting ourselves to only two possible modal qualifications, the biotic and the ethical, would have the following implications. Using the biological modality, represented by the function of saving lives, would, for example, be likely to result in the aiding the dying process being seen as an outrage, going against the very purpose of hospital activities. It would probably also result in hospitals where very sick and weak patients might not be allowed visits from their relatives - as it might compromise their precarious health - and where patients would be forced to stay in the hospital as long as this was considered necessary to sustaining life functions.

If, instead, the ethical modality was chosen as the qualifying function, assistance in dying might in some circumstances be considered more humane and caring. It is also possible that these hospitals would give a higher priority to the social needs of patients, even at the expense of the biological benefit. This could result in less rigid rules regarding visitation privileges and perhaps even permission for patients to be cared for in their homes rather than in the hospital. This line of thinking could lead to reconsideration of where the hospitals are located: close to communities served or where the air, temperature, and humidity are especially favourable. We could also ask: how the hospital should be designed - e.g., how important is it to have playgrounds and playrooms on site, as well as family rooms enabling close relatives to stay overnight or even live with the patient for short or long periods of time? Besides the biological and ethical modality, of course, other dimensions - such as the economical - could also be considered. When discussing different design alternatives, Dooyeweerdian philosophy also proposes that good design is characterized by a vision where all of the modalities are allowed space and where there exists a balance and harmony between the different modalities, thus tempering extremist thinking.

The third application area is the use of the concept to provide a common base for design. Ackoff (1993) argues that, in general, people tend to disagree less about ideals than about shorter-range goals and ways of pursuing them - i.e., "the more ultimate the values, the more agreement they generate." He continues by saying that once agreement has been reached on ultimate values, differences over means and short-range goals can often be resolved. Following this argument, I argue that the concept of qualifying function can be used in order to find accommodation and, hence, has the potential of providing a common base for design, while at the same time allowing for a high degree of variety within the boundary of that common base. Besides being useful for finding common ground between participants, the modalities can also point to fundamental difference of opinion as to what should guide and give meaning to the system in question.

In order to test the efficacy of the three application areas suggested above, the concept will be placed within an existing methodology for system design and then applied to a case study in order to test its practical applicability. Soft Systems Methodology (SSM) will be used as the framework for this discussion, since this methodology recognises and emphasises design as a mental activity which builds upon participants' perceptions of reality. It accounts for their worldviews, as well as how these perceptions and worldviews affect design situations (Checkland, 1981; Checkland and Scholes, 1999).

3.   SOFT SYSTEMS METHODOLOGY

Soft Systems Methodology, SSM, (Checkland, 1981; Checkland and Scholes, 1999) is a well-known methodology that aims at tackling real world problems. As such, it has been used for information systems design (Checkland, 1984, 1988; Checkland and Griffin, 1970; Checkland and Holwell, 1997; Rose, 2000; Wilson, 1992; Winter, Brown and Checkland, 1995) as well as system design and learning in general (Callo and Packham, 1999; Checkland and Scholes, 1990a; Kartowisastro and Kijima, 1994; Mirijamdotter, 1998; Reid, Gray, Kelly and Kemp, 1999; Rose, 1997). It focuses on models of perceptions, not models of complex reality (Checkland, 1981, 1982), and is especially suited for problem situations labelled as complex and pluralistic (Flood and Jackson, 1991a, 1991b). Hence, the focus of attention for SSM is people's perceptions of reality, including their worldview, rather than on external reality as such. This focus on perceptions and worldviews has its rationale in the belief that change in worldview (*Weltanschauung*), precedes significant and sustainable changes in socio-technological systems.

Soft Systems Methodology aims to extend the arena in which systems ideas are used to improve upon - or solve - real world problems in social systems. It has been recognized as, arguably, "the most self-conscious (and certainly the most rigorous) attempt at an interpretive systems methodology" (Jackson, 1982:22). This distinction is achieved by taking seriously the subjectivity which underlies human affairs. SSM guides practioners to "treat this subjectivity, if not exactly scientifically, at least in a way characterized by intellectual rigour" (Checkland, 1981:30). This intention or attribute is visible throughout the methodology and based on the assumption that individuals interpret situations differently and in accordance to what they find meaningful. Further, what is perceived as meaningful is based on our particular 'image' (Boulding, 1956) of the world. This image is, in turn, formed by individual preferences, background, knowledge and experience, which, in general, is tacit knowledge taken entirely for granted. Soft Systems Methodology teases out such world-images or '*Weltanschauungen*' (Checkland, 1981:18) and examines their implications. Because of this, the concept is considered to be among the most important in the methodology.

The importance of the concept *Weltanschauung* and its central position within SSM have also been advanced by other authors (Checkland and Davies, 1986, Fairtlough, 1982, Jackson, 1982, Jayaratna, 1994). This fundamental SSM concept is understood as the set of assumptions taken as given in communication between people. Therefore, when understood, it can help an observer understand social situations. Further, Checkland (1981) argues that whether we realise it or not, we view 'raw' data via a particular mental framework, world-view or *Weltanschauung* and we attribute meaning to the observed activity by interpreting it within the context of our larger framework . Even more significantly, we only view the observed activity as meaningful in terms of a particular *Weltanschauung*, which is invisible to us.

Checkland and Davies (1986) distinguish three categories of the term: W1, W2 and W3. W1 is a taken-as-given set of assumptions that make a particular transformation or root definition meaningful and is there only to help in model building. It has nothing to do with the dynamic flux of events that represents social reality and should be stated as purely and as simply as possible. W2 is related to a version of the problem situation, and thus, related to the taken-as-given assumptions in W1 in the sense that W2 makes W1 relevant. Finally, W3 is of broader concern and related to the social reality in which the problem situation is embedded. W3 is linked to our beliefs and assumptions about reality and makes us understand social situations.

This means, that while W1 is related to the systems thinking and modelling phase, W2 and W3 are not, they are both related to the problem situation existing in the real world. Because of this, neither W2 nor W3 is included in the modelling techniques developed for SSM (Checkland and Davies, 1986, Fairtlough, 1982). Checkland and Davies also say that the concept *Weltanschauung* should be reserved for W2 and W3, while W1 can be referred to as simply W. I understand and agree with the distinction made by Checkland and Davies, but argue that W2 and W3 should be included in the SSM process to a larger extent. By mainly focusing on W1 much of the richness and understanding related to the concept *Weltanschauung* will be put at risk.

However, in order to make the complex notion of *Weltanschauung* manageable, the phenomenon has been reduced quite drastically within the model building phase of SSM. Instead of including the whole *Weltanschauung* of a person, Checkland suggests that it is enough to try and draw out the underlying rational of why a person finds it meaningful to carry out a certain activity or transformation; i.e. W1. Even if this is most helpful, Bergvall-Kåreborn and Grahn (1996b) have argued that the concept has been too narrowly restricted in practice to realize the full potential of different perceptions as well as the diversity present in real life design situations.

In the following a short description of the methodological process of SSM is presented so the reader can perceive how the interpretative element of the methodology is operationalised. The process consists of four main phases: finding out; systems modelling; comparison; and, taking action, as illustrated in Fig. 2 (after Checkland and Holwell, 1993, Checkland and Scholes, 1990a:7). It is not the phases in themselves that makes SSM valuable because, as Checkland (1981) himself points out, they are all 'everyday mental acts' (p.214) and can be found in most methods and methodologies aimed at change. Rather, it is the way these phases are perceived and carried out, due to the epistemological and methodological *Weltanschauung* of the originators of SSM.

Even though it is possible to start the process at any phase - i.e., it is the relation between the phases rather then their order that is important) it usually begins with the first phase by an exploration of a real-world situation of concern (the left centre of the figure), initiated because someone perceives that situation as problematic and wants to do something about it. The purpose of the exploration is to provide a better understanding of the situation in question and it is usually summarised in a so-called 'rich picture' (Checkland and Scholes, 1990a).

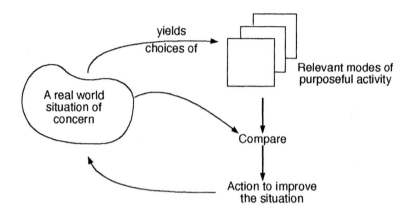

Figure 2. The basic shape of SSM

From the rich picture, issues, judged by the analyst or someone else to be relevant for improving the problem situation, are selected and modelled using system concepts. These models, depicted as square boxes in the upper right-hand corner of Fig. 2, are intellectual devices used to stimulate and structure debate about the problem situation under study. They illustrate different perceptions or interpretations of the real-world situation under study and represent activities that logically need to be performed in order to reach a certain purpose. Because of this, they are refereed to as conceptual models of 'human activity systems' (HAS) (Checkland, 1971, Checkland, 1981, Checkland and Scholes, 1990a), and the class of human activity systems comprises all activities that are carried out by human beings. In order to form a 'whole' or a system, these activities are linked to each other by some principle of coherence or some underlying purpose or mission. Further, these models should be understood as neither accounts of the real world nor Utopian designs, but rather epistemological devices which help to structure a debate.

In the third phase, the models of human activity systems are set against actual perceptions of the situation, based on individuals' appreciative settings and, to some extent, depicted in the rich picture (the right centre of the figure). Through the comparison, and the debate it creates, new insights are revealed and appreciative settings may be changed, hopefully in such a way that accommodations between different interests and views can be reached. Accommodations that emerge as both feasible and desirable lead the way towards actions to improve the situation. This represents the fourth phase. It is depicted at the bottom of the figure. The implementation of agreed upon changes, or actions to improve the situation, then becomes the new problem situation. In this way, the methodology comes full circle. However, if changes cannot be agreed upon, a more extended examination of relevant systems will be necessary, initiating a cyclical review process.

The purpose of the systems thinking part, or modelling phase, is to draw out different perspectives of the problem situation and to structure the thinking of the same. In order to do this, some precise techniques have been developed, consisting of root definition (RD), CATWOE, PQR, and conceptual models (CM) of human activity systems (HAS). Root definition means naming, in a short statement, a system of purposeful activity. The formal rules for a well-formulated root definition is that it should contain the elements of the mnemonic word CATWOE (Smyth and Checkland, 1976) or PQR[1] (Checkland, 1999, Checkland and Scholes, 1990). PQR refers to the statement, "Do P by Q in order to contribute to achieving R", and answers the three questions: What to do (P); How to do it (Q); and Why do it (R)? (Checkland, 1999:A23), while CATWOE stands for: Customers; Actors; Transformation; Weltanschauung; Owners; and Environmental constraints.

4. THE CASE STUDY

In order to test the practical applicability of the concept of qualifying function as a tool for system design, it was embedded within Soft Systems Methodology and applied to a case study. The study focused on two EU programs, Women In Progress (WIP) and AdAstra, both of which were located in a small municipality in the north inland of Sweden. They aimed at creating new work opportunities by helping people with business ideas to develop and operationalize these. Hence, self-employment in some form was a key concept.

In order to help people develop and operationalize their ideas, the program offered a thirty week education comprised of two main parts. The first part was a more or less common ten week education, covering topics such as information technology (IT), accounting, human communication, and presentation techniques, as well as business idea development. During these ten weeks, the students were also expected to further their own ideas

---

[1] In earlier writings PQR has always been referred to as XYZ, but in Checkland (1999) it is changed with the argument that Y might be confused with *why* in the question "*Why* carry out the transformation?"

as business plans. After the initial ten weeks, participants had a follow-up period of up to twenty weeks of individual operationalization work. The students where selected on the basis of personal interviews. A total of three succeeding groups of thirty persons each participated in the projects.

Even though WIP and AdAstra formally constituted two separate projects with separate owners, target groups, goals and financial arrangements, the programs ran seamlessly as one intervention program with the same project leader and reference group. The reasons for this decision were systemic: the projects had common purposes and common methods for accomplishment. Besides this, or perhaps influencing this, was the fact that both projects had been initiated and developed by the office for Trade and Industry within the municipality. On the other hand, the projects focused on different target groups. The target group for WIP was unemployed women with low education and a long previous employment within the public sector. AdAstra was directed towards people currently employed within the public sector who ran the risk of loosing their jobs. This difference disallowed the Office for Trade and Industry from representing both projects, because unemployed people lie outside their commission.

Using the powerful statement of SSM's PQR (Checkland and Scholes, 1999) in order to summarise my first understanding of the projects from a formal perspective results in the following representation:

What to do (P)?:	Develop models of how to create new work opportunities
How to do it (Q)?:	Through projects like WIP/AdAstra
Why do it (R)?:	In order to reduce the intermediate stage between unemployment and work (WIP) and help companies and organisations to better prepare for structural changes in society (AdAstra)

5.   RESEARCH METHODOLOGY

Appreciation of the case study requires further context so I next describe the research methodology guiding my research. My research follows the tradition of the interpretative and action oriented approach whereby the researcher enters a real-world situation with the aim of both improving it and acquiring knowledge (Baskerville and Wood-Harper, 1998, 1996; Checkland and Holwell, 1998; Hult and Lennung, 1980; Susman, 1983). Action research approaches are becoming more and more accepted and adopted within the research tradition of systems theory and information systems as the benefits of this approach are disclosed (Baskerville and Wood-Harper, 1996; Checkland and Holwell, 1998; Galliers and Land, 1987). When it come to research regarding systems development methodologies Baskerville and Wood-Harper (1996) even argue that "action research is one of the few valid research approaches that researchers can legitimately employ to study the effects of specific alterations in systems development methodologies. It is both rigorous and relevant" (p.240). However, despite this strong recommendation, there is still some confusion as to what should count as action research (Baskerville and Wood-Harper, 1998; Seashore, 1976; Warmington, 1980). My work is best described by the action research type developed by Checkland in SSM (Checkland and Holwell, 1997, 1998).

My role or involvement in the study was of a facilitative nature, which means that while the responsibility of solving the problem rested with the study subjects, my task was to clarify how different project members viewed the projects and by that increase their understanding of the projects as well as their role in them. This was particularly important, since the two projects were to be run as one. Also, the project leader recognized varying views among project owners and project participants, which needed to be reconciled in order for the collaboration to succeed.

It was decided that I would be the one to carry out the study - e.g. to draw the rich picture and develop the conceptual models - while the project members would contribute information about their perspectives of the projects. The study would culminate in a workshop in which I would present a rich picture of the situation and selected models of relevant systems to be discussed by the different project members. This event was to precipitate the negotiation of worldviews necessary to bridging differences and forging new alliances for cooperative work.

Further, following this type of action research it is particularly important to discuss the concepts, framework of ideas, methodology and area of application, although Checkland (1985, 1991; Checkland and Holwell, 1997) argues that these three concepts need to be defined, discussed and related to each other no matter which research method one uses. My framework includes systems thinking, as viewed by the soft approach together with Dooyeweerdian philosophy. As my methodology I will use Soft Systems Methodology supplemented with the modal theory and the concept of qualifying function from the philosophy of Dooyeweerd in order to facilitate the process of generating different perspectives and conceptual models of two EU projects focused on creating new work opportunities (A). Both A and M are further explained below.

In order to gain an understanding of the situation in question, I used interviews as my main source, conducted as a mix of semi-structured interviews and open discussions. In total, I interviewed thirteen persons: six students, a tutor, the project leader, both project owners, and three persons from the steering committee and reference group.

6.   APPLYING THE CONCEPT IN A PRACTICAL STUDY

During the finding out phase, many interesting ideas for relevant human activity systems emerged (Bergvall-Kåreborn, 2002a), such as:

♦ A system to create new work opportunities,
♦ A system to help people take responsibility for their development and future,
♦ A system to take advantage of people's resources,
♦ A system to help people develop themselves,
♦ A system to change and remodel the structure of trade and industry.

The finding out phase also confirmed the project leader's feeling that different key persons within the two projects had different views on the projects. Amongst the most important differences in perspective were disagreements on the aim of the projects as well as on how the projects were to be run. Two important parts that from the outset were perceived as common for the two projects, and influenced the decision to merge the projects on an operational level.

The owner of WIP, a private organisation, Institut Ungdom och Framtid, IUF, (Institute Youth and Future), saw the projects as a 30-week long project consisting of three phases in which all students participated. The first phase, in their minds, dealt with self-development, group work and communication. In this phase, the students were to make an inventory of present and future needs in their local community and analyse the present supply. Based on this, the students should try to find their own business idea and formulate a successful business plan. Phase two would focus on how to take responsibility for a business. At this point, the students would have an opportunity to try and start a business. In phase three, the students should have their business up and running, but would still have guidance and supervision from the tutors and project leaders, as well as from the group as a whole. The only reason for a student not to follow the 30-week program would be if the person after the first phase did not have any clear ideas of what to do and how to do it.

The owner of AdAstra, the office for Trade and Industry within the municipality, viewed the projects as an individual journey consisting of two phases. The first phase was a 10-week general educational period that was quite similar for most students, but that could be adapted to fit particular needs as well. After this phase, an individual program should be developed for each student, clarifying what the student wanted to achieve and how this was to be achieved, as well as important stages and dates along the way. No student should be involved in the projects longer than needed to realise his or her goals.

In order to test how the concept of qualifying function together with the modal framework could be used to elicit different perspectives, I used the modalities to further understand what could lie behind the strive to *create new work opportunities through self-employment.* To explore possible views on this, I used statements from the participants relating to their view on the purpose of the projects, together with the modalities, as stimuli. This revealed that some of the statements - such as *"utilise the resources available in people"* - felt closely related to a certain modality (economics), while other statements - *"help people find their place in society"* - could easily be related to several modalities (ethical, social, economic) and shifted their meaning depending on which modality was chosen for interpretation. For instance, in interpreting the statement, *"help people find their place in society"*, from an ethical aspect, students became the focus. As the main customers, their needs, interests and well-being guided the creation of work opportunities. In viewing the statement from the social modality, however, the main customer became society (state or municipality) and the focus for creating work became directed towards the social structure and where in this structure the different students would fit best. This perspective also made visible work as a socialising activity, where (young) people's norms, values and practices are fostered. Work became an activity that both educates people and gives them a role and place in society. Relating the statement to the economic modality, society was still the prime customer but the focus for the projects became much more geared towards resource use than social structure.

In the following discussion, a number of possible views will be given on what qualifies or makes the activity of creating new work opportunities meaningful. These views either are based solely on a modality or constitute a mixture of a participant's statement and a modality. In the case of the latter, this will be indicated by including the statement within brackets at the end of the paragraph.

Starting with the *ethical* modality again as the qualifying function, but relating it to a different statement, one could again argue that there are tasks, like taking care of the sick and the old, that need to be completed, and human kindness and consideration demand that we fulfil them. These tasks also need to be carried out if we want our societies to be humane and caring. Hence, according to this perspective, the main function of creating work is to be able to provide for those that are not able to care for themselves, at least not fully. This view can be related to the statement, *"turn needs into work opportunities"*.

Viewing create new work opportunities from a *juridical* qualifying function rather than an ethical puts focus on human rights, rather than human kindness and consideration. From this perspective, it could be argued that in order for justice to be served, everybody needs to work in order to contribute to, and help retain or increase social welfare. Hence, working or not working is not a personal choice but the duty of every man and woman. Related to this line of argument, but seen from another angle, is the view that the purpose of work is to distribute the wealth (resources), as well as the responsibilities, of a society. This line of thinking suggests that, since it is through work that a society's resources are distributed, it is the right of every man and woman, rather then the duty, to have the possibility to work. This second viewpoint is, to some extent, observable in countries where the state creates and pays for opportunities to help unemployed people get back into the labour market, as well as to keep the official numbers of unemployed down.

Seen from an *economic* perspective, and given that different people have varying gifts and potentials, the prime function of organised work is to take advantage of and utilise these potentials (competence, experience, knowledge, ideas, enthusiasm, etc.) in a wealth producing fashion. Not doing so would be a waste of resources. This would also be the rationale for work specialisation and other related concepts and can be illustrated by the very simple example of two neighbours trading work assignments with each other. One, trained as a bricklayer, might help her neighbour to install a tiled stove, while the neighbour - a skilled landscape architect - helps her design her garden in return. Seen in a bigger perspective, this exchange would be perceived as a way for people to divide the work tasks in society in order to improve the quality of work and/or reduce the time needed to carry out the tasks. This view can be associated to the statement, *"to utilize people's resources and ideas"*.

Another perspective on work, still seen from the economic aspect, is that its main function is to create revenue and increase the buying power of people. So, from this economic perspective, the focus would be on demand (needs) and supply. Ways would be sought to satisfy demands with as little resource use as possible. While the statement, *"turn needs into work opportunities"*, was interpreted from an ethical perspective above, it has a different meaning when interpreted from this second economic perspective. This is an example of where the same statement, interpreted from different perspectives, generates different meanings.

Finally, we can view work from an *historical* perspective, guided by the kernel of formative power, and by that representing the past as well as the future. Seen this way, how and where work is carried out has a far-reaching effect when it comes to shaping a society as a whole, such an effect that one could argue that this is its main function. Viewing work according to these, or any other perspective, places the activity of creating work opportunities in new contexts. The statement, *"change and remodel the structure for trade and industry"*, can represent this final view.

6.1 Two conceptual models based on different qualifying functions

Below a minimum number of logically contingent activities that need to be carried out in order to accomplish what has been defined in the root definition - "A system to create new work opportunities through self-employment because it is seen as good for the municipality" - are presented in two conceptual models (CM) (Checkland, 1979b; Checkland and Scholes, 1990b). The models are based on the same PQR statement but related to different qualifying functions. The first CM is based on the ethical modality and represents the view that new work opportunities are good for a municipality because they help the municipality address inherent needs and because it gives people a possibility to create jobs that feel meaningful to them. The second CM is based on the historical modality and represents the view that new work opportunities are good for a municipality because they can help change and remodel the structure for trade and industry. The latter refers to a structural change which aims at reducing the municipality's dependency of the public sector as a main employer for its inhabitancies by increasing the number of work opportunities within the private sector. The reason for my choice of these two functions is that they to some extent represent the views of key persons in each project.

While both conceptual models aim to create new work opportunities, the underlying rational for doing so is different and this is reflected in the models' activities. In Fig. 3, which has the ethical modality as the qualifying function, the focus for the model is to create work opportunities in order to provide for people who are not able to care for themselves and at the same time help people develop jobs that feel meaningful to them.

Here, it is the perceived needs of both the receivers and providers that guide what kind of work that should be created. In Fig. 4, where the conceptual model is drawn from an historical perspective, the focus is on creating work opportunities that have the possibility to change and remodel the regional structure of trade and industry. Hence, the different qualifying functions highlight different types of customers (C in CATWOE) the ethical function focus on customers on an individual level while the historical centres on customers at a societal level.

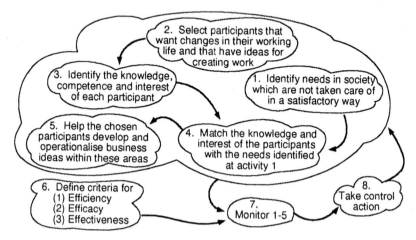

Figure 3. Conceptual model for how work opportunities that benefit a municipality can be created seen from an ethical perspective

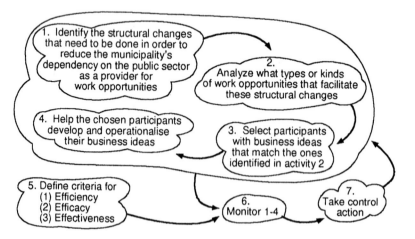

Figure 4. Conceptual model for how work opportunities that benefit a municipality can be created seen from a historical perspective

This analysis is in line with Mingers' (1990) argument concerning the what/how distinction related to conceptual models where he argues that there can be situations in which the how is more important that the *what*. "There may be agreement about what is to be done but disagreement about how to do it. This can be modelled either by having one RD and a series of CMs, or in preference, a number of RDs which only differ in respect of the *how* which is specified in them" (p.24). In the case of WIP/AdAstra, there was a general agreement as to what should be done - i.e., create work opportunities - but the reasons for this and agreement on how this would be accomplished turned out to be more problematic. The way forward in this study was to develop a number of RDs which only differed in respect of the *why* which was specified with the help of the concept of qualifying function and affected the question of *how*.

After taking a broad view on the issue at hand, the concept was also used to facilitate the process of generating different views on what makes it meaningful to try and develop models and methods for how new work opportunities can be created. Based on the following P and Q, different views on R or W1 were elicited with the help of the modal framework.

What to do (P)?:   Develop models or methods of how new work opportunities can be created

How to do it (Q)?:   Through projects like WIP and AdAstra

Why do it (R)?:   ◆ it is good resource management, and hence economic, to develop models that other people can use and reuse (economic)
                     ◆ we can and should learn from each others past experiences (historical)
                     ◆ the Swedish society is similar enough to other societies within the European Union for the knowledge from the projects (WIP and AdAstra) to be transferred, acceptable and usable in other countries within the EU (social)
                     ◆ it is possible to communicate the knowledge gained through the projects in the form of models and methods to people not involved (lingual)

According to the SSM literature, W (of CATWOE) is said to represent a worldview which makes the transformation meaningful (Smyth and Checkland, 1976). Looking through case studies (Checkland and Scholes, 1999), however, reveals that W (or R in the case of PQR) is often used in two different ways: (1) to illustrate different perspectives on why it is meaningful (desirable) to carry out the transformation and (2) why it is seen as possible (feasible) to carry out the transformation. This difference is illustrated in the four perspectives listed above. While the first two reasons are related to desirability, the latter two are related to feasibility. Both ways of using W contributes to the modelling process by pointing at underlying assumption of different types. To make the most of this difference, it is, however, important to make clear how W is used in a particular root definition.

At the end of the study, based on SSM and modal insights, it was decided by the project owners, in consultancy with the project leader, that the student groups should be split into two groups after the basic 10-week education. One group consisted of people who needed more personal development. They were to receive guidance of a more social character. The second group were people who had a fairly clear business idea and needed help on developing and operationalise it. Even though it was mostly WIP students in the first group and AdAstra students in the second group, the split was not made according to project membership but, rather, according to the assessed needs of the individuals. The split also made it easier for each of the project owners to pursue their underlying purpose with the project they were in charge of.

7.   LEARNING

The aim of this chapter was to present an intellectual tool that facilitates the process of unfolding underlying rationalities related to problem definitions and improvement suggestions in system design, with a special focus on the methodology developed by Checkland.

Based on the experience gained in the case study, I argue that the concept of qualifying function can complement existing techniques - such as SSM's root definition, CATWOE or PQR, and conceptual model - for drawing out different perspectives of a problem situation, i.e. modal thinking helps to elicit different kinds of *Weltanschauungen* on different levels. In the study, this was first done by using the concept, together with the modal framework, as a stimulus for exploring possible views on what qualifies, or makes the activity of creating new work opportunities meaningful. This focused attention on perspectives related to the problem situation as such (W2), as well as beliefs and assumptions about the social reality in which the problem situation was embedded (W3). Hence, it broadened the perspective on *Weltanschauung* used in the traditional SSM modelling phase. Including W2 and W3 in the modelling proved to have very positive consequences for the study as a whole since it was on this level that the more fundamental differences between the participants appeared. One obvious implication was the difference between creating work in order to provide for those that are not able to care for themselves as contrasted with creating work in order to shape and re-model society as a whole.

The qualifying function was also used as a stimulus to elicit different W1s related to why it is seen as both desirable and feasible to develop models for how work can be created. However, even with this purpose in mind, the concept had a tendency to lift the thought of *Weltanschauung* to the level of W2 and W3 rather than W1, resulting in reasons such as: "it is good resource management to develop models that other people can use and reuse" or "it is possible to communicate the knowledge gained through the projects in the form of models and methods to people not involved".

The notion of a qualifying concept also helped to clarify the meaning of a stated *Weltanschauung* (W) or Transformation (T) - and to show that many different interpretations can be made of a certain statement. In this study, this clarification was achieved by relating different modalities to statements about T and/or W in order to explicate them before writing the root definition. Most of the time, a statement produced quite different meanings depending on the modality to which it was related. When exploring possible views on why creating work opportunities can be seen as good for a municipality, for instance, the statement, "help people find their place in society", clearly illustrated this point. The meaning of the statement shifted appreciably depending on whether it was interpreted from an ethical, social, or economic perspective. Though the examples in this study was drawn entirely from the modelling phase, it is also possible to achieve this level of clarification in the discussion phase.

Incorporating the concept of qualifying function into the root definition by clarifying the Transformation (T) and/or *Weltanschauung* (W) also introduces a new way to strengthen "the [fragile] 'bond' between a conceptual model and a root definition" (Checkland and Tsouvalis, 1997) and thus strengthen the defensibility of the conceptual model. Defensibility, here, refers to the second of the two validity questions posed by Checkland (1995) to answer how well one can tell a 'good' design from a 'bad' one and focus on "whether a given model is competently built" (p.52ff). It does not, however, mean that only one CM may be drawn from any given RD as explicated by Checkland and Tsouvalis (1997), Mingers (1990), and Schregenberger (1982). Rather, the concept provides an additional bond between RD and CM that might provide an alternative to the technique for checking the technical defensibility of a conceptual model. It can, for example, complement the techniques of using the 'Formal Systems Model' (Checkland, 1981), modelling the transformation process of the root definition in such a way that it fulfils the criteria of effectiveness, efficiency, and efficacy (Checkland and Scholes, 1990a), or including all the main activities of the conceptual model in the root definition (Checkland and Tsouvalis, 1997).

Further, clarifying T and/or W influences the model building in that it added knowledge to the root definition, which in turn helped to find relevant activities that enrich the conceptual models. This helped to make the transition from root definition to conceptual model. Enriching the conceptual models also address the repeated problem concerning the tendency of the conceptual models to be impersonal and general and by that not always very informative (Checkland and Scholes, 1999; Checkland and Tsouvalis, 1997; Mingers, 1992; Mirijamdotter, 1998; Naughton, 1979; Schregenberger, 1982). To develop models that are personal and relevant for participants is of outmost importance since the "models in SSM have an impact on and effects in the real world, affecting perceptions of the problem situation" (Tsouvalis and Checkland, 1996).

Finally, during this study only about half of the modalities played an active role in stimulating the modelling and only a handful reoccurred on a regular bases. Common for the more actively used modalities are their relation to the social sciences (ethics, economics, history) while modalities related to the natural sciences very seldom have been used. The reasons for why and when different modalities are particularly useful can most certainly be explained by a mixture of issues such as the context, problem situation, change process phase, and user interests. However, in order for a defining explanation of pattern to emerge, many studies need to be carried out. Besides this, most of the stated Ws relate to the pistic modality in that they are all beliefs. This is also in agreement with earlier findings by Bergvall-Kåreborn and Grahn (1996b) in their discussion regarding the relation between the pistic modality and the concept of *Weltanschauung*. However, while the statements can be related to the pistic modality, I was not able to find what felt like relevant W1 specifically based on the pistic modality.

8. CONCLUSION

This chapter argues the importance of distinguishing and negotiating different perspectives or discourses at individual and group levels, especially in organisations that cross traditional boundaries. But before we can learn from this diversity we first need to communicate and share our perspectives with each other. Consequently, there is a need for conceptual tools that can help us to express and explicate this diversity. This chapter presents one such tool, the concept of qualifying function and its related modal framework, embedded in Soft Systems Methodology. The chapter illustrates how the concept can be used to facilitate the process of unfolding underlying rationalities related to problem definitions and improvement suggestions in system design. The presented findings are based on both theoretical and practical studies and relates to SSM as well as to system design situations in general.

Firstly, the concept of qualifying function proved a useful tool for drawing out different perspectives of a problem situation and for identifying self-imposed constraints. As such, it can complement existing techniques (such as SSM's root definition, CATWOE or PQR, and conceptual model). In the study, the concept, together with the modal framework, was used as a stimulus for exploring possible views on what qualifies, or makes the activity of creating new work opportunities meaningful. This generated a number of interesting perspectives on why a municipality might view this activity as important.

The study indicates that the qualifying function has a tendency to focus on perceptions related to the problem situation and to beliefs and assumptions about the social reality in which the problem situation is embedded, rather than to the *Weltanschauung* of a particular transformation. Broadening the perspective on *Weltanschauung* used in the modelling phase also broadened the understanding of the relevant system under study, the primary-task system for creating new work opportunities. This proved to have very positive consequences for the study as a whole since it was on this level that the more fundamental differences between the participants appeared.

Secondly, the concept of qualifying function helped to clarify the meaning of a stated *Weltanschauung* (W) or Transformation (T) and to show that many different interpretations can be made of a certain statement. In this study, this clarification was done by relating different modalities to statements about T and/or W in order to explicate these before writing the root definition. Most of the time, a statement got quite different meanings depending on which modality it was related to. Thus, within SSM the qualifying function provided support for clarifying and enriching PQR and CATWOE statements.

Thirdly, clarifying T and/or W influenced the model building in that it added knowledge to the root definition, which, in turn, helped to find relevant activities that enriched the conceptual models. This facilitated the transition from root definition to conceptual model. As such, it is one way to help inexperienced model-builders to make the transition from root definition to conceptual model.

Fourthly, the concept of qualifying function also relates the root definition and the conceptual model closer together and this way provides an alternative technique for model validation.

Finally, the concept proved to be a useful modelling tool for drawing out different perspectives of a particular problem situation, discussing desirable purposes for the design, and for exploring the underlying rationale behind a suggested transformation or a stated *Weltanschauung*. The concept of qualifying function, together with the modal framework, helped the participants to view their situation from new and different perspectives. Through this, the study resulted in a change of attitude or outlook for the project members by increasing their understanding both of the projects as such, as well as their role in them. This shared understanding would have provided a useful starting point for the cooperation between the projects, and most certainly influenced the approach and the extent of the cooperation.

Chapter 4

ELICITATION OF HIGHLY INTERDISCIPLINARY KNOWLEDGE

Mike Winfield and Andrew Basden

1. INTRODUCTION

Disciplines involve knowledge, some that can be articulated and shared, and some that is tacit and not easily shared. Interdisciplinarity involves a sharing of knowledge between various disciplines. When interdisciplinarity is of the wider, integrative type (Boden, 1999; see chapter 1) there is a need to develop genuine mutual shared understanding between disciplines as widely separated as physics, jurisprudence, sociology, economics and theology.

In recent years, those working in the arena of knowledge management have recognised the widely interdisciplinary nature of knowledge shared within an organisation and have made much use of Polanyi's (1967) notion of tacit knowledge. The value of tacit knowledge is, to Yates-Mercer and Bawden (2002:22) at least, that it is "holistic, involving an appreciation of things in their totality: Polanyi refers to this as an 'indwelling'."

But what is this holism, this 'totality'? And how can we elicit the totality of knowledge that is relevant to an interdisciplinary situation? These are important questions when we make knowledge available to others, especially throughout an organization. This paper discusses them.

2. MULTIPLE FACETS OF KNOWLEDGE

"Attempts to define knowledge," say Kakabadse, Kouzmin and Kakabadse (2001:141), "reflect the multifaceted nature of knowledge itself." By using this metaphor, of facets of a jewel cut along its distinct planes of separation that cut across each other at angles that are given by the nature of the material itself, they are emphasizing that there are distinct areas of knowledge, but also that the areas, as it were, cut across each other and relate to each other in ways that do not jeopardize their independence. What we might think of as the holistic totality of the beauty of the jewel comes not from any single facet, but when all the facets are well developed and all play their part.

While discussing knowledge of individuals versus collective knowledge Lang (2001:46) states, "knowledge produced locally does not have explicit exchange value", suggesting that the knowledge must be placed in a wider context. She suggests that a knowledge worker will need to acquire new knowledge every four or five years but this knowledge will, due to the changes that most profoundly affect a body of knowledge, have to come from other domains. Hence she is arguing very strongly that a multi-disciplinary and multi-aspectual view of knowledge is necessary.

As Earl (2001:218) points out, in contrast to "the expert systems idealized in the 1980's which were too inflexible and narrow", there is a need to capture many facets of knowledge, even, he points out, in the relatively constrained tasks of chemical engineering, where "such knowledge bases tend to be domain-specific, supporting and improving specific knowledge-intensive work tasks and particular sorts of decision-making. ... This knowledge base was used to analyze chemical behaviours, process practices, product parameters, and environmental conditions in order to optimise factory performance." Even in this limited example, we see human and organizational practice being relevant alongside chemical behaviours and environmental conditions. Johannessen, Olaisen and Olsen (2001) note that a company that does not emphasize the entire knowledge base will lose competitive advantage.

But the multifaceted nature of knowledge is even more pronounced when it supports wider business tasks (Jacob and Ebrahimpur, 2001:78):

> "The majority of the managers have a long history in the company and successful leadership of projects is dependent on long tenure since the preference is for managers with a generalist competence profile. In the words of one interviewee: One needs to have as broad a knowledge base as possible. It is the outer parameters that one must have knowledge about."

Though loosely defined, 'outer parameters' indicates facets that are often overlooked. In interdisciplinary knowledge 'outer parameters' abound.

2.1   Elicitation of multifaceted knowledge

Just as the whole beauty of a jewel depends on every facet being operational and all contributing together, so the success of a system depends on every facet of knowledge playing its part. The omission of (knowledge about) any facet can jeopardize the whole. As Yates-Mercer and Bawden (2002:22) say, "An attempt to focus on the 'particulars', the detailed and tangible make-up of the thing, destroys what Polanyi regards as the 'truer' tacit appreciation." If interdisciplinary knowledge is to be useful within an organisation and across its departments it must be elicited and made available in a form that can be shared. Therefore all facets need to be elicited, and elicited in such a way that the relationships between them are clearly understood. This leads to two main questions:

1. How do we identify all the facets of the knowledge, and especially those 'outer parameters' that are often overlooked?
2. How do we relate disparate facets to each other?

2.2 Existing methods of eliciting interdisciplinary knowledge

Cognitive mapping (Eden, 1988) is a largely graphical technique by which participants set down concepts, make their meaning clear by the method of contrast, and then link them together into a map. Much use is made of the associations we make between one concept and another. This technique has proved very useful for identifying which of the concepts in a complex, detailed situation are the key ones.

But in widely interdisciplinary situations it is important to ensure that 'outer parameters' are not overlooked. Cognitive mapping has no way of ensuring that oft-overlooked aspects of a situation are brought into the picture. This implies that the process of knowledge elicitation must be designed to stimulate the participants in analysis to think of these 'outer' aspects that are too often overlooked. One way is to make use of the notion of the participants' perspectives or '*Weltanschauungen*'.

Soft systems methodology (SSM) (Checkland, 1981) sees the concept of *Weltanschauung* (or 'W') as the most important one in the methodology (p.18). It is a well-developed methodology comprising a number of stages or analysing the problem situation to gain a rich picture, then abstracting away from it to model possible system that could bring about useful transformations. During the modelling stages, participants are recommended to focus on six elements (customer, actors, transformation, W, owner, environmental contstraints: CATWOE). SSM is widely used, but has been subjected to criticism (e.g. chapter 5), one of which is that it is not good at handling conflicts that arise from differing Ws (Jackson, 1991). Though a range of Ws are examined during analysis, these are simply those held by participants, so there is a danger of groupthink or that some affected stakeholders are not participants in the analysis (such as animals and the environment). SSM does not disturb "structural features of social reality" (Jackson, 1991:164).

Strategic Assumptions Surfacing and Testing (SAST) (Mason and Mitroff, 1981) can be very useful for making Ws explicit and turns conflict into a virtue. It proceeds by four stages. First, people representing a wide range of perspectives are assembled into groups, within which viewpoints converge but between which the divergence of view is maximized. Within their groups, participants unearth their most significant assumptions, usually using "Who?" questions. Stage three is dialectic debate, in which each group defends their preferred strategy and states their key assumptions, and each group attempts to understand the strategies and assumptions of other groups. Stage four, integration, involves coming to agreement about the overall key assumptions and negotiation between groups regarding strategy. If no agreement is reached, points of disagreement are noted and research is carried out to resolve them.

SAST could, therefore, be used to bring diverse and disparate Ws to the fore and thus might be more successful than SSM in uncovering the roots of conflict and in surfacing a wider number of perspectives. It is characterized as: adversarial, participative, integrative and 'managerial mind supporting'.

But, again, there is no guarantee that all 'outer parameters' are likely to be uncovered, nor that the mutual understanding is reached about all the 'facets'. The reliance on working in groups might lead to participants defending their views and preferred strategies and being unwilling to open up and consider alternatives. The emphasis, in stage two, on "Who?" questions can exacerbate this. The adversarial nature of SAST might cause problems for some, such as those exhibiting Asperger's Syndrome (Attwood, 2001) who find it closes their thinking down. It might also shift the group towards assuming that each person will hold only one perspective, and thus the richness of multiple perspectives can be lost. The dialectic process in stage three seems to presuppose Habermasian conditions of ideal dialogue, in which the force of the better argument wins, and that all participants are willing to take the trouble to understand the views and strategies of others; in many situations this presupposition does not fully hold.

Therefore, we still seek a way of understanding all the 'facets' of knowledge and for bringing to light all the 'outer parameters' so often overlooked.

3. A PHILOSOPHICAL UNDERSTANDING OF THE MULTI-ASPECTUAL NATURE OF KNOWLEDGE

3.1 Why philosophy?

To answer such questions requires more than research: it requires a philosophical examination of our framework for understanding what knowledge is, what is involved in knowing (or, as Polanyi refers to it, 'appreciation'), and what it is we have knowledge of. The criteria for, and notion of, truth in physics, sociology and theology (for example) are all radically different, and from within each science other notions of truth seem irrelevant or worse. As the Cambridge Encyclopaedia of Philosophy points out, science tells us what is true, while philosophy tells us what truth is. It is philosophy, therefore, rather than science or logic, that enables us to recognise and respect the diverse notions of what is held to be true, valid or rational in the various disciplines - to recognise the different 'facets' of knowledge. It is philosophy that enables us to bring these 'facets' together.

So, if we are to have genuine shared understanding between disciplines we need to approach interdisciplinarity with philosophical rather than scientific tools and methods.

Moreover, what is tacitly known by us is often of our lifeworld, that stock of shared understandings which, Habermas (quoted in Ray, 1993:31) says, "dissolves ... before our eyes as soon as we try to take it up piece by piece." In the extreme, perhaps, tacit knowledge cannot be articulated or shared but in less extreme cases, some articulation may be possible. Many of the 'outer parameters' are not completely inarticulable but merely overlooked and taken for granted. We must address the relationship between tacit knowing and what we might make explicit when we explore it. Again, it is philosophy, not science nor logic, that concerns itself with the relationship between the lifeworld and articulated, theoretical knowledge.

This paper discusses an attempt to employ a particular strand of philosophical thought, to address both the philosophical questions of totality and indwelling, and the practical question of how to elicit it, first worked out by Winfield (2000).

The philosophy we employ is that of Dooyeweerd (1955), who is one of the few philosophers to have addressed these questions in sufficient depth. While Kant critically addressed the questions of how we observe and how we reason, Dooyeweerd critically addressed the deeper question of what it means to take a theoretical attitude as such, as distinct from the everyday attitude taken in the lifeworld. While Habermas (1972) sought to draw attention to the human interests that inescapably pervade knowledge and the related disciplines, and then to discourse by which validity claims could be challenged (1984), Dooyeweerd sought to understand the nature of such interests and the basis on which validity claims could even be meaningful in the first place.

As discussed in chapter 2, Dooyeweerd started with everyday life and its diversity of meaning and yet its coherence. This led him into a theory of modal aspects which is perhaps the most visible part of his philosophy today.

3.2   Aspects

Dooyeweerd held that there is a law-side to reality, a framework of law that enables all that happens in the cosmos and all that exists, which is distinct from our concrete experience, which he called the subject or entity side. The law side is a framework in which things can be meaningful and by which things exist, occur and are guided - in which we 'dwell'. Thus it is not unlike Polanyi's notion of 'indwelling' (Yates-Mercer and Bawden, 2002) and not unlike Heidegger's notion of 'worlding'.

But, Dooyeweerd noted, we experience considerable diversity and also an element of coherence as we 'dwell'. He accounted for this diversity and coherence by proposing there are distinct spheres of meaning, which he called modal aspects or law-spheres. For example being a vetinary surgeon has a biotic aspect and an economic aspect - as well juridical, social, ethical and other aspects. The entire suite of aspects that Dooyeweerd delineated is:

- ♦ Pistic aspect, of faith, commitment and vision
- ♦ Ethical aspect, of self-giving love, generosity, care
- ♦ Juridical aspect, of 'what is due', rights, responsibilities
- ♦ Aesthetic aspect, of harmony, surprise and fun
- ♦ Economic aspect, of frugality, skilled use of limited resources
- ♦ Social aspect, of social interaction, relationships and institutions
- ♦ Lingual aspect, of symbolic meaning and communication
- ♦ Formative aspect, of history, culture, creativity, achievement, technology
- ♦ Analytical aspect, of distinction, conceptualizing and inferring
- ♦ Sensitive aspect, of sense, feeling and emotion
- ♦ Biotic (organic) aspect, of life functions, integrity of organism
- ♦ Physical aspect, of energy and mass
- ♦ Kinematic aspect, of flowing movement

♦ Spatial aspect, of continuous extension
♦ Quantitative aspect, of amount.

The aspects towards the bottom of this list (Dooyeweerd called them 'earlier' aspects) tend to be determinative while those at the top ('later') tend to be normative. Note that the list is generic and not just relevant to our example (vets); different disciplines draw from the same list, though perhaps with a different pattern of emphasis.

The meaning of each aspect is ultimately beyond precise human understanding, and is better grasped by intuition rather than by theoretical thought or by definition because these aspects exhibit an "indissoluble inner coherence" (Dooyeweerd, 1955,I:3) that makes lifeworld possible as a 'totality'.

In general, every aspect is important in human activity. This is true not only of everyday life but also of professional life when it becomes mature, practised expertise. This means that if any aspect is ignored then that activity might be jeopardised. For example, vetinary surgeons who reduce their practice to the economic and ignore the ethical or social aspects are likely to sacrifice customer loyalty and face disciplinary action from the Royal College, while those who reduce practice to the biotic and ignore the economic are likely to soon go out of business.

Dooyeweerd accounted for this by saying that no aspect can be reduced to another, and that when we try to do so, then we rob that aspect of its meaning. We cannot rely on good functioning in one aspect automatically being accompanied by good functioning in another (nor vice versa). As the reader may have realised, the poetic notion of facets may be founded philosophically on Dooyeweerd's notion of aspects.

3.3   Aspectual perspectives

Each aspect, being a sphere of meaning, provides a way of taking a distinct perspective from which to view any thing, event or situation. When we conceptualize a situation, we usually do so in terms of concepts meaningful in a certain aspect. Assumptions we make are usually related to specific aspects. For example, vetinary practice can be viewed from a biotic perspective, in which case such concepts as vitality, blood pressure, muscle tone are seen to be meaningful without explicit thought. From a legal perspective, concepts like ownership, contracts and fair dealing are meaningful, and it might be assumed that the customer who brings an animal for inspection is the owner. Each aspect yields a different notion of "what counts as valid and worthwhile evidence or knowledge" (Lang, 2001).

We can make valid inferences among the concepts and propositions that are meaningful within a given aspect. But, because of the irreducibility between aspects, inferences cannot be validly made that cross aspectual boundaries, and there is no transcending logic by which we can reason ourselves into a different aspect to ensure that all aspects are covered. Therefore we must consider each aspect in its own right. The 'outer parameters' that Jacob and Ebrahimpur (2001) speak of could be concepts from an aspect we are not currently focusing upon.

In interdisciplinary situations, Lang (2001:45) says, "knowledge is created and recreated as practitioners see the logic of each other's thinking in communities brought together by common interests." But Nosek and McNeese (1997:118) note "the difficulty and rarity of actually thinking of the question from another discipline's perspective." Fahey and Prusak (1998:268) speak of "individuals' differing perspectives, beliefs, assumptions, and views of the future" that "are most likely to collide and thus immobilize decision making." The use of the word 'collide' implies something more fundamental than a mere disagreement over facts or deductions; it refers to a complete lack of mutual comprehension that sometimes exists between parties, and an inability to see that the other perspective has any relevance at all. The challenge of wide interdisciplinarity is that what is meaningful in one aspect is viewed as being relatively meaningless in or to another, and has a tendency to be oversimplified or ignored by it, and this can lead to conflict.

3.4   Disciplines

Each aspect provides a distinct way of knowing a distinct set of meaningful things to study and know. Each aspect therefore lends itself to be the centre of a competence, discipline or science which is developed in "communities with their own evaluative practices" (Lang, 2001:46); e.g. the juridical aspect is the core of what legal disciplines study and practice, just as the physical aspect is the core of what physicists, chemists and materials scientists study and practice. When Augier and Vendelø (1999:253) say that "within a firm one can find islands of specialized knowledge" this can refer to more than specialization in depth, it can also refer to the distinct scientific or disciplinary areas centred on aspects. Some disciplines related to each aspect are as follows:

- ♦ Pistic aspect: theology, religious leadership, vision-building
- ♦ Ethical aspect: ethics, the charity sector
- ♦ Juridical aspect: jurisprudence, politics
- ♦ Aesthetic aspect: art, music, film, leisure, sport

♦ Economic aspect: economics, management
♦ Social aspect: institutional studies, sociology
♦ Lingual aspect: linguistics, semiotics, languages
♦ Formative aspect: history, culture, technology
♦ Analytical aspect: logic, analysis
♦ Sensitive aspect: psychology, psychiatry
♦ Biotic (organic) aspect: biology, medicine
♦ Physical aspect: physics, chemistry, materials science
♦ Kinematic aspect: mechanics, animation
♦ Spatial aspect: geography, topology, geometry
♦ Quantitative aspect: statistics, arithmetic.

3.5   Interdisciplinary shared understanding

Augier and Vendelø continue, "Such knowledge needs to be combined or cross-fertilized with knowledge from other sub-units to stay viable and valuable to the firm and perhaps even more important prevent sub-units from getting caught in competence traps." But it is seldom sufficient merely to bring different specialists together into a team and hope the result will be interdisciplinary. That each aspect possesses a distinct rationality means that if each participant in the team proffers only their formal or specialised knowledge, there is little hope of such cross-fertilization. However, this may be possible if we take note of the multi-aspectuality of human activity.

A discipline is not merely an island of pure rationality, however, but part of human living that may be seen as lifeworld. For example, the work of vetinary surgeons, chemists and priests (which might centre on the biotic, physical and pistic aspects respectively) also has a social aspect, as they deal with people, a juridical aspect, as they both give what is due to others and respect the law of the land and the bylaws of their profession, and so on.

Dooyeweerd held that all human living in fact involves all aspects. What differentiates one discipline from another lies in the aspects that are most important to it and 'qualify' it (the notion of qualifying aspect is discussed by Bergvall-Kåreborn in chapter 3).

Therefore each practitioner, whatever their discipline, is likely to have at least some awareness of the meaning of most aspects, even if they focus on only one. This gives us a possible foundation for mutual, shared understanding in a widely interdisciplinary team. If we are to analyse the multiple aspects of a situation from an interdisciplinary standpoint, we must find a way of maintaining holistic awareness of all aspects even while we might focus on one of them. Since there is no automatic logical connection between concepts that are meaningful in one aspect to those meaningful in another, we must find a way to explicitly attend to each and every aspect that is relevant.

4.   MAKE: MULTI-ASPECTUAL KNOWLEDGE ELICITATION

We need a framework which transcends both the participants in the analysis of a situation and the situation being analysed, so that it contains at least an element that is not dependent on the perspectives of the participants. Dooyeweerd's aspects provide such a framework.

One obvious method for analysing interdisciplinary situations is to use the suite of aspects as a checklist, presenting each aspect in turn. But this would force the participants into focusing too narrowly on analytic ways of knowing and would most likely miss the other important ways of knowing. It would not capitalise on the participant's associative links.

Though aspects are irreducible to each other, Dooyeweerd also maintained there are relationships among the aspects, of two types. Later aspects can depend on earlier ones, and each aspect contains 'echoes' of all the others. This enables us to form meaningful associations between concepts whose meaning is found in different aspects, and facilitates our analysis of situations. Instead of seeking rational links, we seek those associations that result from inter-aspect dependency and analogy.

The Multi-Aspectual Knowledge Elicitation method (MAKE) is a simple method devised by Winfield (2000) for achieving good aspectual awareness of a complex situation in a more natural way. It capitalises both on the irreducibility of aspects, by means of the participant's intuitive grasp of the meaning of individual aspects, and on the inter-aspect relationships, via the participants' associative links. We explain it briefly here, prior to discussing its ability to explicate tacit knowledge.

4.1   The method of MAKE

First, the notion of aspects is explained to the participants (usually one interviewee), and their kernel meanings are explained in simple terms in such a way that an intuitive grasp may be readily formed.

The overall approach is to start by asking the experts in the application domain to identify a couple of the aspects they deem most important, and to grow a recognition of the relevance of other aspects by a gradual process via associations. The participants first identify a few whole aspects, then they start to identify concepts and laws etc. within aspects.

In the process, concepts come to mind that do not fit well within currently identified aspects, and the participants are thus led in a very natural manner to identify other relevant aspects. The steps of the MAKE process may be seen as:

1. Introduction (e.g. explanation of kernel meanings of aspects, and obtain statement of requirements, or some other entry point)
2. Identify a few (e.g. a couple) important aspects.
3. Focus on one of these aspects and specify any laws, axioms, data, definitions, and constraints that apply to the domain.
4. Identify as many concepts as possible that lie in this aspect. (Note: May need to check the concepts at a later stage to ensure they fall within the correct aspect.)
5. Apply low level abstraction to each concept, which needs, or is thought to need exploding.
6. Repeat steps 3-6 as necessary.
7. Use the aspectual template to identify any new aspects, which may apply to the concepts specified but (build bridges between concepts and aspects), and return to step 3.

Almost always a diagram of the aspects and their concepts is drawn as the discussion proceeds, to make something reminiscent of cognitive maps but in which concepts are grouped together in aspects that look like clouds. There are relationships between concepts, between aspects and between aspects and concepts. See Fig. 1 for a simple aspect map based on Winfield, Basden and Cresswell (1995). The maps are drawn as the interview progresses, with the interviewee being involved in drawing. It is useful if the first map drawn is given to the interviewee to think over and see if they wish to change anything. Typically the interviews are up to an hour in duration, but it can be carried out in several stages.

Winfield refined and tested MAKE on eight case studies (Winfield, 2000), mostly with participants who had never heard of Dooyeweerd or the aspects before, and found consistently that MAKE was easy to learn (both by interviewees and by Winfield's students who carried out some of the case studies) it was not difficult for participants to grasp enough understanding of the aspects in order to undertake this process, that nearly every aspect (typically 13 out of 15) was identified in each of the case studies, that it is useful over a wide variety of domains (tree planting, sustainability, vetinary practice, Islamic food laws, youth advice and management of a local housing business unit).

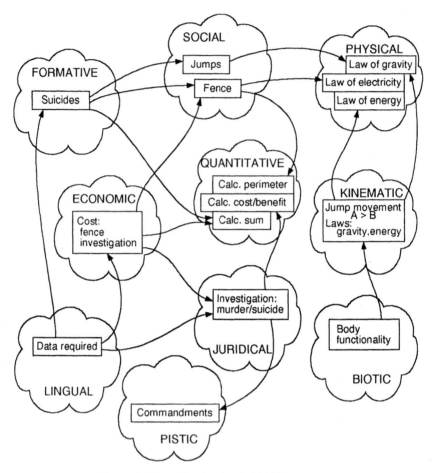

Figure 1. Example of a simple MAKE aspect map

4.2   Discussion of MAKE

MAKE has a flexible structure that places the interviewee very much at the heart of the knowledge elicitation process. From the eight cases mentioned by Winfield (2000), together with sundry other cases since then, MAKE has been found to have the following characteristics.

4.2.1   On the explication of knowledge

♦ MAKE unearths concepts and relationships, within spheres of meaning, rather than rules. However, some of the relationships can represent rules.

♦ MAKE operates by helping people to make connections across aspects that are meaningful to them.

♦ In most cases a variety of aspects appeared quickly and easily. MAKE is good at stimulating people to what might be called lateral thinking, to uncovering things that are usually taken for granted or not considered.

♦ MAKE is successful in helping to draw attention to aspects outside the usual focus of attention so that 'outer parameters' are not overlooked. Most interviewees find that almost every aspect is relevant (but without being forced to do so; see below). While MAKE does not elicit detailed tacit knowledge, it seems extremely efficient at eliciting the kind of tacit knowledge that is simply taken for granted because it is not of the central aspects of the interviewee's main activity.

4.2.2　On the surfacing of assumptions and Weltanschauungen

♦ Because of its ability to make oft-overlooked concepts explicit, it has the effect of bringing assumptions to the surface. Though the focus is on unearthing concepts rather than assumptions about them, this effect has been noticed. MAKE could be adapted to bring out assumptions as such.

♦ Because it makes the participants aware of each aspect, MAKE can be used specifically for the making Weltanschauungen visible. It has the advantage over SSM of stimulating participants to look beyond the perspectives held by the group. It has several advantages over SAST. In being used by individuals not in a group setting, there is greater openness and less tendency to defend positions; the individual is stimulated not by other individuals but by the transcending suite of aspects. For similar reasons, it is non-adversarial. Moreover, MAKE does not depend as heavily as SAST does on the quality of dialogue, nor on goodwill among partipants - though ironically it encourages goodwill because it draws attention to the ethical aspect of self-giving.

♦ Unlike SAST and to some extent SSM MAKE helps to evoke multiple perspectives within a single individual, because it focuses not on the 'who' but on the 'why'.

♦ If MAKE is undertaken with several participants, and their maps shared, it provides the basis for a richer mutual understanding and respect. Suppose one participant finds the concepts of another irrelevant at first. By drawing attention to the aspect in which they are meaningful, the first participant can come to acknowledge their relevance. Focusing on aspects de-emphasises differences of personality or organisational role, so conflict is reduced, and the views of non-dominant participants is more readily accepted.

4.2.3   On the process

♦ We can detect a hermeneutic circle in MAKE, insofar as the analysis proceeds in cycles that consider whole aspects and detailed concepts in turn. It is the ability to move naturally between these that leads to concepts emerging that belong to different aspects.

♦ The aspects were very readily understood by all the interviewees. The fact that many were 'ordinary people', rather than academics, is interesting. This supports Dooyeweerd's claim that the kernel meanings of the aspects are grasped by the intuition rather than by theoretical thought, a claim that has also been supported by Lombardi (2001).

♦ Because the introduction to the aspects is usually brief, participants will often have a distorted or incomplete understanding of their meaning. First, this is recognised and it can be useful to cast their meanings in phrases that will be salient to the interviewee. Second, some degree of misunderstanding does not seem to matter because both interviewer and interviewee operate with intuitive rather than precise understandings. The kernel meaning of an aspect can sometimes appear in the constellation of concepts attributed to the aspect in spite of some misunderstanding. This effect might support Dooyeweerd's claim that the kernel meanings are grasped by the intuition and can never be bounded by precise definition.

♦ Some suggest that using a ready-made suite of aapects would inhibit the interview or distort the content that emerged, but this does not seem to be the case. Surprisingly, MAKE is both liberating and empowering (if used well), as in the following.

♦ Liberating. Neve (2003:48) argues that when eliciting knowledge through interviewing the questions should "promote individual awareness and mutual comprehension and not lock the individual, nor mislead, or give incorrect association." When MAKE was used it was, almost universally, the interviewee who identified the aspects, and not the interviewer. The questions the interviewer asked were simple ones like "Why?" and "What else?"; seldom were leading questions asked about a particular aspect, and then only towards the end after most of the information had been obtained. Thus MAKE seems to provide considerable freedom for the interviewee to express what is meaningful to them; this arises from the aspects being spheres of broad meaning, and we touch on it below.

♦ Empowering. There have been occasions when MAKE and reference to the aspects seemed to empower the interviewee, to say things that they felt slightly uncomfortable or embarrassed about. We attribute this empowering effect to the fact that the aspects constitute a framework that transcends the interviewee's situation which they can 'hold on to'.

♦ MAKE helps to promote interdisciplinary communication. Through the use of the maps prepared by different communities looking at the problem from different views, and then comparing the maps to enable discussion and reflection upon differences and similarities. Identifying the aspects in which concepts are meaningful, helps to separate real differences from those created by different naming conventions.

♦ Would MAKE be compatible with Glaser and Strauss's (1967) Grounded Theory? At first sight, being guided by a set of pre-agreed aspects might seem to be the antithesis of Grounded Theory, in which theories are supposed to arise from the data itself and not from any prior concept-structure imposed by the researcher. However, we must not forget the difference between aspects and concept-structures. The aspects are not, themselves, concept structures, but are broad spheres of meaning that enable participants to make meaningful distinctions within the situation under study and thus form their own concept structures.

5. CONCLUSION

We have shown how taking Dooyeweerd's theory of irreducible yet intertwined aspects of meaning as a starting point we can generate a method for analysis of ill-structured, interdisciplinary domains of knowledge and work. Aspects are a law-framework in which we 'dwell', and indeed make such dwelling possible. Dooyeweerd's suite of aspects was an attempt at complete coverage of types of meaning. While this aim will never be fully realized (Dooyeweerd, 1955,II:556), they provide a useful framework to guide analysis of the holistic totality of an interdisciplinary situation.

MAKE, Multi-Aspectual Knowledge Elicitation, is a simple iterative method based on these aspects, which is easy to understand and apply, and seems friendly to the interviewees, who have found Dooyeweerd's aspects intuitive. In this, it exhibits some of the characteristics of cognitive mapping, but, by using aspects to guide the process, it also exhibits a sensitivity to world views similar to SSM. Like SAST it enables participants to surface assumptions, and yet can elicit more detailed knowledge in the context of aspectual assumptions, and has proved successful in eliciting many 'facets' and bringing to light the 'outer parameters'. It combines the identification of aspects of an application with an initial more detailed analysis of the individual issues, concepts, laws, etc. that are important for the application, and with a means of relating both aspects and concepts.

MAKE seems to liberate and empower the interviewee rather than constraining them to a pre-given framework, because it provides not a framework derived from theory so much as one composed of broad ways in which things in life can be meaningful. This gives it a refreshing quality and it is being explored in further research and practice.

Chapter 5

AN APPRECIATIVE CRITIQUE AND REFINEMENT OF CHECKLAND'S SOFT SYSTEMS METHODOLOGY

Anita Mirijamdotter and Birgitta Bergvall-Kåreborn

1. INTRODUCTION

Contemporary organisations experience continuous and fast paced change due to, amongst other things, developments in information and communication technology (ICT). The technology enables interconnectedness, global communication and information exchange. This, in turn, leads to new artificially created environments, accelerated access to information, and exposure to cultural diversity. It also leads to increased competition among businesses which stimulates development but limits the planning horizon. Due to the interconnectedness and the increased uncertainties that such situations create, organizations need to act dynamically, developing their abilities to adapt flexibly, responsively, and innovatively (Holst, Mirijamdotter, Bergvall-Kåreborn and Oskarsson, 2004). This has profound consequences for ideal organizational redesign (Mirijamdotter and Somerville, 2004; Somerville and Vazquez, 2004; Somerville, Huston and Mirijamdotter, 2005). One way to design more dynamic organisations is to organize work in project teams (e.g. Senge, 1990; Dahlbom, 2000), including cross functional groups (Holst, 2003, 2004; Holst and Mirijamdotter, 2004a, 2004b). If properly planned/designed, in these situations, synergistic interactions occur across multiple, distributed work groups, oftentimes on a global scale.

The interconnectedness within and among dynamic organizations suggests that our societies - and, in fact, the whole of the universe - can be perceived as a system (chapter 6). Moreover, the process of enquiring about contemporary organisations can successfully be organized as a system (Checkland, 2000b; Checkland and Scholes 1990a). This was discovered by the University of Lancaster researchers in their attempt to apply systems engineering methodology to social and managerial problems (Checkland, 1972, 1981; Wilson, 1990). Early on, they recognized the limitations of systems engineering 'means-end' thinking, which could not define 'messy real world' problems nor report multiple perspectives on problematic situations. The researchers also acknowledged the need to include human values and invite stakeholders' ideas. The methodology that evolved was termed Soft Systems Methodology or SSM (Checkland, 1981, 2000b; Checkland and Scholes, 1990a; Wilson 1990, 2001). Two of its main features are, in the analysis phase, the so-called Rich Pictures and, in the design phase, the Human Activity Systems. These information visualization techniques capture and communicate stakeholders' conceptions of and perceptions about real world problem situations.

Over three decades of SSM usage has proven its advantages over the means-end thinking, represented by systems engineering, in contexts of uncertainty and plurality. The inclusion of human values and the representation of systems concepts have been especially important. However, today's challenges - created by the rapid and continuous change in which modern organizations operate - require further enrichment of SSM. For instance, in today's workplace environments, projects are often completed by distributed work groups. Task completion requires formal and informal 'sensemaking' (Weick, 1995) frameworks, held in common among individuals and across teams. In such situations, incorporation of compatible concepts and frameworks can further the transformative enablement of Soft Systems Methodology. In this chapter, for instance, we illustrate the enrichment potential of incorporating Dooyeweerdian philosophy (Kalsbeek, 1975; Dooyeweerd, 1997). The result, enhanced understanding of stakeholders' points of view, can facilitate increased desire for and pursuit of 'ideal' (Weber, 1949) systems among workplace participants. Enrichment of selected Soft Systems Methodology techniques with Dooyeweerdian thinking demonstrates the situated benefits and efficacy of a coherent enhanced framework.

We begin with presentation of selected concepts from Soft Systems Methodology and Dooyeweerdian philosophy. Starting with discussion of SSM from a systems design point of view, we focus on finding out about the problem situation through gathering information for creation of a Rich Picture. Then, we explain the modelling methodology for Human Activity Systems and focus on the subject of evaluation criteria. Thereafter, we turn our attention to Dooyeweerdian philosophy and focus on the modal aspects and qualifying, or leading, function. This section is followed by explorations of incorporating parts of Dooyeweerdian philosophy, first into the process of gathering Rich Pictures and then into valuing models of Human Activity Systems. The chapter concludes by summing up the benefits of this extended methodology and finally reflections on research-in-progress in which we seek to embed the approach in organizational design. In so doing, we intend to test the efficacy of this enriched SSM approach to ensuring continuous organizational learning adequate to anticipating appropriate organizational responses.

2.   SOFT SYSTEMS METHODOLOGY

Soft Systems Methodology "is a methodology which recognizes the role of the individual's 'world images' and the influence of historical background on the interpretation of 'reality'" (Jayaratna, 1994:176). Though SSM has its roots in Systems Analysis and Systems Engineering, it has steadily moved away from the principles characterizing the so-called hard methodologies. Its basic shape (after Checkland and Scholes, 1990a:7, Checkland and Holwell, 1993) is shown in Fig. 2 of chapter 3.

The SSM process usually starts because someone perceives a real-world situation as problematic and wants to do something about it. In order to obtain a better understanding of the situation of concern, relevant issues are illustrated in models of purposeful activity. These models reveal different interpretations of the real-world situation under study and represent activities that logically need to be performed in order to reach agreed upon purpose(s). The models are then compared with what is perceived, i.e., the actual perception of the situation. Through the comparison, new insights are revealed which will lead to clarification of purpose and corresponding actions to improve the situation.

In the process of identifying relevant issues, the SSM practitioner gathers information about the problem situation itself and represents that information. This representation is termed 'Rich Pictures' (Checkland, 1981:317; Checkland and Scholes, 1990a:45). However, it is important to note that the term Rich Picture does not only include the pictorial representation itself, but the whole process of appreciating the situation (Lewis, 1992). The technique for representing the problem situation builds on the recommendation to look for "elements of slow-to-change structure" and "elements of continuously-changing process" and how these elements relate to each other within the situation climate (Checkland, 1981:164).

The notion of climate has been further developed to include explorations of the situation through intervention analysis, social system analysis, and political system analysis, i.e., SSM's Analysis One, Two, and Three (Checkland and Scholes, 1990a). As is typical in systems work, the emphasis is on representing relationships and connections. Therefore, a diagrammatic representation - rather than linear text - better expresses these relationships. So although the Rich Picture has come to be associated with this diagram, as discussed above, Lewis (1992:358) points out that the diagram is rather "a by-product of the process of investigation of the problem situation" - i.e., it is reflective of the outcome of purposeful inquiry.

In a complementary modelling technique in which purposeful activity, or Human Activity Systems (Checkland, 1979b; Checkland and Scholes, 1990a), are represented, a variety of stakeholders' perspectives involved in the 'problem situation' (Checkland, 1981:155) are put forward to interpret the 'real' world. In this systems thinking part, some 'precise' techniques have been developed, including Root Definition, CATWOE, PQR, and Conceptual Models of activity systems. Root Definition means naming, in a short statement, a system of purposeful activity. Experience has shown the importance of observing formal rules for a well-formulated root definition, e.g., that it should contain the elements of the mnemonic word CATWOE (Smyth and Checkland, 1976): Customer, Actor, Transformation, Weltanschauung, Owner and Environmental constraints. Later, Checkland condensed this framework into a shorter version as represented by the letter sequence PQR (Checkland, 1999:A23) which stands for: Do P by Q in order to achieve R. CATWOE and PQR have a central role in modelling because they bring forth various perspectives, including diverse participants' tacit assumptions.

The CATWOE technique has been examined by Bergvall-Kåreborn, Mirijamdotter and Basden (2004) in which they suggest replacing some of the terms: Affectees for Customer, Process for Transformation, Decision taker for Owner and Constraints for Environmental Constraints, to meet some of the difficulties with the original terms experienced especially by novice users. They suggest the mnemonic PAWDAC which represents the elements Process, Actor, Weltanschauung, Decision taker, Affectees and Constraints. Our action research to date suggests that these new naming conventions more easily cultivate individual and group understanding and thereby facilitates improved level and quality of participation. However, since the purpose of this chapter is the enrichment of SSM through incorporation of Dooyeweerdian thinking, we retain the original elements in this text.

Another dimension of SSM modelling is the diagrammatic versions of the process stated in the root definition, known as Conceptual Models. These visual models consist of a set of logically contingent activities, supplemented with a monitor and controlling sub-system that evaluates the Human Activity System's performance. The criteria for evaluation, normally referred to as the 3 (or 5) E's (Checkland, Forbes and Martin, 1990), include:

E1 - Efficacy   Are these activities accomplishing the stated output of the process? Does the means work?

E2 - Efficiency   Are minimum resources used? Could the process be carried out with fewer resources using a different technique?

E3 - Effectiveness Are we doing the right thing? Are we accomplishing our longer-term goals and are they aligned with the weltanschauung/worldview of the owner?

E4 - Ethics   Is this a morally correct thing to do?

E5 - Elegance   Is the process aesthetically pleasing?

These measures of performance are meant to judge the "in-principle performance of a human activity system" (Checkland and Scholes, 1990a:42) and are, as such, focused on the process of the modelled system. The first three measures constitute the base and originate from the question "How could the system fail?" (Checkland, 1989:90). Ethics and elegance are enrichments which can be neglected, replaced, or complemented by others of free choice.

As depicted in its basic shape (Fig. 2 of chapter 3), the next SSM process phase is Comparison. In this phase a variety of stakeholders' perspectives are debated with a focus on exploring the system. We find that this debate is often occurring 'already' in the modelling phase, when modelling Human Activity Systems and identifying relevant measures for the system's process. The debate is intended to lead to the Actions stage which changes the perceptions of the problem situation and thereby initiates a new SSM cycle.

Intentional learning is a natural outcome of SSM practice. When SSM thinking informs an inquiring process, the whole methodology becomes a tool for intentional learning and purposeful change. This is significant because 'dynamic organisations' (Holst et al., 2004) require ongoing reflective and responsive changes that further adaption to changing circumstances. However, a study of SSM use among non-academics

(Mingers and Taylor, 1992) found that the majority of participants chose SSM to develop an understanding of the situation, not to bring about change. In response, we seek to develop a more robust version of SSM which more explicitly advances the individual and organization value of using SSM for both individual and group insight and also for directed change. To optimize the learning potential of SSM and to further enrich the understanding of stakeholders' points of view, we suggest the incorporation of a Dooyeweerdian philosophy framework, which provides a coherent applied theory for attributing meaning to particular contexts, in intuitive terms based on human experience and thought.

3.   DOOYEWEERDIAN PHILOSOPHY AND ITS FRAMEWORK OF DIMENSIONS

More recently, scholars in Information Systems have turned their attention to the work of Dooyeweerd, e.g. (Basden, 2002; Bergvall-Kåreborn, Mirijamdotter and Basden 2004; De Raadt, 1995; Eriksson, 2001; Winfield, Basden, and Cresswell, 1996) and introduced additional aspects, besides logics and economics, into systems analysis and design. The intention of this section is to introduce the modal aspects of Dooyeweerdian philosophy in preparation for then exploring the possibilities for their interaction with SSM.

When studying political situations, Dooyeweerd discovered that several distinct dimensions of human experience and thought relate to spheres of life. These dimensions, organised in a holistic framework ranging from determinative to normative, are: numeric, spatial, kinematic, physical, biotic, sensitive, logical, historic, informatory or lingual, social, economic, aesthetic, juridical, ethical and credal. See Fig. 1.

It is important to note that the distinctions between the modal aspects are made for purely analytical reasons. In human life and experience, they are bounded together as a whole and one dimension cannot replace another. On the other hand, the ordering of the aspects are not haphazard; on the contrary, there are interrelationships between the dimensions which define their position. Thus, the aesthetic aspect is dependent on the economic, the economic on the social, the social on the lingual and so on, and no one dimension can be reduced or replaced by any of the others. We have chosen to represent the modal framework in a fan to illustrate that the framework will loose its capacity as the number of modalities is ignored in the same way as the fan will loose its capacity as the number of feathers is reduced. We use the same metaphor for organizational design -

i.e., if one aspect is missing, work and work place may not suffer, but if several aspects are lost, or ignored, the functioning of the organization will in the long run erode.

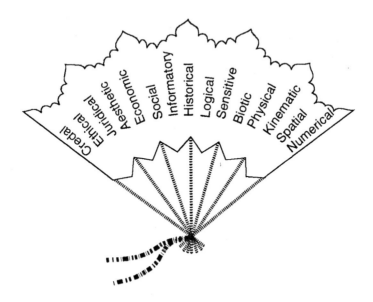

Figure 1. The different modal aspects and their interrelation

Each modal aspect is given its meaning through its kernel, or nucleus, and has its own order, or set of laws, by which it is governed. Thus, the numeric aspect is guided by its kernel, a discrete quantity, which is different from symbolic representation, the kernel of the lingual aspect. The modal aspects and their kernels are best illustrated by an example and, for that purpose, we illustrate their application in relation to the activity of designing an Information System (IS). A more elaborated example can be found in Bergvall-Kåreborn (2002b).

When designing an information system, the numerical aspect is essential since it is the foundation for information technology. Further, discrete quantity is exemplified in number of designers, users as well as time and money. The spatial aspect refers to issues such as the physical, mental or social space of the organisation that the IS and its users may occupy. Some work within organisations requires people to move around large areas at times. Because of this the mobility of the technology is an important issue to consider in relation to the kinematic perspective, which illustrates motion. The physical dimension focuses on issues such as energy provision for the system.

When designing an information system the biotic aspect points to the fact that users have different biological conditions. Some might have reduced sight or hearing, others might be disabled. The sensitive perspective creates a sensitivity to feelings, both positive and negative, that the change process and the implementation of the system cause.

The logical aspect points to distinctive differences in needs, knowledge, involvement, responsibility and authority between stakeholders. It also illustrates the distinction between different types of flows, such as logical flows and information flows. The historical perspective deals with the present and the future, formed by the past. A design situation is always contextualised by a certain history, a history of which there will always be more than one account. Good design requires learning and reflecting on this history.

Information is embodied and communicated with the help of symbols, operating in the lingual dimension, which represent and model the roles, structures, processes, and flows comprising an IS project. The design of an information system can be seen as a social activity that affects social relationships (roles, norms and power relations) and by that creates a new social milieu. The economic aspect is often used as a major justification for an IS investment. In business it is often perceived as maximising profit while Dooyeweerd's notion represents a long time perspective that focuses on economising resource use. The aesthetic aspect is a criterion that is becoming more and more important in the design of technological devices, such as computers and mobile phones. However, the aspect does not solely deal with beauty, it also points to the importance of harmonising subsystems, such as the technical and the social systems.

An IS project usually includes many juridical agreements that direct and control the activity and regulate relations between the involved parties. The ethical aspect refers to discussions about what we can and have the right to do, on one hand, and what we feel we ought to and want to do, on the other hand. It calls attention to the responsibility of individual stakeholders. Finally, the credal perspective represents our basic attitudes and value systems. While some might believe that information technology is the answer to all problems, others might believe that this technology undermines the competence of professionals and, as such, represents a threat to modern society.

Just as each dimension has a kernel that informs its meaningfulness, every designed system should be defined by a core aspect which guides and gives meaning to the system's purpose. This aspect defines the core purpose of the system and is, according to De Raadt's terminology

(1991:110), named the system's 'sphere of sovereignty'. We also refer to it as the leading, or qualifying function, which is discussed in more depth in chapter 3. While a certain kernel is linked to a specific aspect, a system's leading function has to be chosen and this modal aspect will affect the activities and the decisions of the whole system.

For example, if an information system's leading function is stated in the economic aspect then the focus of the design is on how the system can reduce costs or increase revenues - or in Dooyeweerd's terms, how the system can improve the resource use. On the other hand, if the leading function is lingual, the intention focuses on the system's ability to mediate relevant data, facilitate communication, and support information and communication technology (ICT) integration into organisational processes.

Finally, Dooyeweerd's dimensions can also be found in our sciences, e.g. the numerical dimension is researched in Mathematics, the spatial in Geometry, the kinematic in Chemistry, while the aesthetical dimension is researched in Arts, the juridical in Law, the ethical in Ethics and the credal in Theology. This scientific distinction promotes the possibility to further our understanding of each dimension as well as, when acting as systems scientists, to identify their regularities and their relations.

As has been noted in the introductory chapter of this book, a narrow, restricted approach to technology in engineering often fails, and technologists themselves are becoming more and more dissatisfied by the limitations of this approach. This is also the situation for conventional design methodologies in the information systems field. Since the nature of our thinking both shapes and limits the analysis and design, we propose that a variety of perspectives need to be included in the systems we create and evaluate.

To illustrate the efficacy of this approach, we begin by demonstrating how these dimensions can be incorporated into the process of appreciating the situation. Next we present a case from a real world situation which shows how Dooyeweerd's framework of modal aspects can support the collection of information for appreciating the situation and thereby enrich SSM's Rich Picture building process. In the subsequent section we present the enrichment potential of incorporating modal aspects into system design and evaluation.

4.   ENRICHING SSM'S RICH PICTURE

In this section, we explore the benefits of supplementing the Finding Out phase of SSM with Dooyeweerdian thinking to create an even more robust methodology. We illustrate the use of modal aspects and leading function when structuring the data gathered for appreciating a problem situation. The data for this case study were gathered to inform local authorities, decision makers and youth social workers about young people's perception of their conditions in a small municipality, Arvidsjaur, in the North of Sweden. The intention was by conscious design to increase future attractiveness of the municipality. Some background information: Arvidsjaur is situated inland and is surrounded by forests, mountains and about 4000 lakes and rivers. By comparison, the number of inhabitants is only around 8000. The major employers were the state and the municipality, especially the army, medical centres and the forestry industry. Due to cuts in governmental expenditures, Arvidsjaur had experienced a growing rate of unemployment. In these respects Arvidsjaur situation was no different from many other inland municipalities in the north of Sweden, i.e., sparsely inhabited with high unemployment rates and dependency upon government employment. What distinguished Arvidsjaur was its strategy to develop work opportunities based in information technology. In order to manage this, an important issue arose, namely how to keep the young people from leaving the municipality for a bigger city and how to encourage the people who leave Arvidsjaur for higher education to return.

Research findings on what was perceived as good and bad about being a youth in Arvidsjaur explored living conditions, cultural activities and leisure time, education, work opportunities, power perceptions, and finally, youths' visions for their future in the municipality. When organising the rich research data, and thus beginning to diagramatically illustrate the Rich Picture, we chose to structure it according to the framework of modal aspects presented earlier. We used the Dooyeweerdian concept of leading function to focus the main issues. In most cases, we selected a leading function that we found to be consistent with the topic of each question or, for the general questions, the answers. In some cases, however, we have chosen to explore different alternatives instead, to illustrate that for certain topics a group of modal aspects are seen as equally important. Below we will give a short version of our process; for a more elaborate version, see Bergvall-Kåreborn and Grahn (1996a).

We started our rich picture drawing by clustering the answers from questions on what was perceived as good and bad about being a youth in Arvidsjaur and then labelled them with the most pertinent modal aspect(s). What was perceived as good we illustrate with stars. The clouds represent what was perceived as bad and the ellipses point to visions for the future. See Fig. 2. We also included the above mentioned pre-conditions of Arvidsjaur (depicted in square boxes) since we believe they were of interest when analysing the answers and also because they made the picture richer.

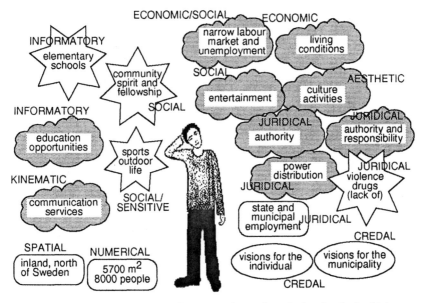

Figure 2. Rich picture representing perceptions of youth situation in Arvidsjaur

Beginning with the square boxes, the municipality's size and the number of inhabitants are linked to the numerical aspect, the bottom left of the figure. Arvidsjaur is situated in the north inland of Sweden (spatial dimension) and, finally, the contract of employment, being foremost a juridical matter, is indicated by history between the inhabitants and mainly the state or the municipality.

The good thing about Arvidsjaur, communicated by the stars, are excellent conditions for sport activities and outdoor life, good community spirit and fellowship, and little violence and drugs. (The term 'drugs' does not include alcohol.) The fact that Arvidsjaur is a small town where almost everybody knows everybody and its location (the numerical and spatial dimensions) relate to the positive things about Arvidsjaur. That was also a common appreciation among the respondents. The last star depicts the perception of good elementary schools in the municipality. This can be related to size and location, but also to the commonality of Swedish elementary school system throughout the country.

The main clouds that appeared in the answers regarding negative conditions for youth in Arvidsjaur were the limited entertainment possibilities (social dimension) and high unemployment (economic and social aspects). The former is a cluster of opinions regarding not enough discos, pubs, cinemas, theatre, concerts, or cafés. This can be related to both the numerical and the spatial aspects mentioned: it is uncommon for a small municipality in the Nordic inland of Sweden to offer significant entertainment possibilities. This by no means indicates that one cannot or should not try to enhance these. The second cloud, unemployment, is foremost an economic and social problem, related to dependency on state and municipal authorities for creating work opportunities. This concept is illustrated in the square box labelled as juridical.

Further clouds appeared when we continued our analyses. Beginning at the lower left, the kinematic aspect was visible in answers relating to several questions. It was perceived that communication services were insufficient and needed to be developed. Continuing in the figure clockwise, from more determinative to more normative dimensions, we return to the informatory aspect. In questions about education, the answers mainly addressed the issue of limited possibilities for higher studies. The fact that the elementary school was perceived as good while there was a dissatisfaction concerning higher education can be explained by a prime difference between the two educational levels; elementary school is quite standardised while higher education is more specialised. Relating to the numerical and spatial dimensions again, it can be hard for a small municipality to offer a wide variety of education.

Besides opinions on too few and too limited entertainment possibilities (illustrated in the social dimensions and discussed in the above) the answers also showed that the cultural activities, which we relate to the aesthetic dimension, were perceived as inadequate. Our data does not permit us to interpret precisely what is meant by cultural activities as Arvidsjaur is an old town with local culture that can be easily accessed by both citizens and local authorities. So in this case size and location are not diminishing factors. However, if cultural activities refer to theatre and operas, then we immediately see difficulties in providing such kind of entertainment in this remote Swedish location.

The question about work opportunities revealed similar answers as the ones related to the first topic concerning good and bad conditions for young people in Arvidsjaur as discussed above. The labour market was perceived as too narrow and the unemployment rate too high. The question regarding living conditions yielded answers which all pointed in the same direction, namely to the economical dimension and to the fact that Arvidsjaur's living conditions were too expensive.

The next topic relates to power perceptions, which we link to the juridical dimension. What was apparent when analysing these answers was that young people felt left out; they felt that no one listened to them or took them seriously; they felt they had no influence on the decisions taken concerning the municipality's future - nor even their own destinies. The power distribution was perceived as unequal. However, also related to the juridical aspect, youth's longing for influence and authority was one sided. By that we mean that we simultaneously found a lack of proactive responsibility for developing life possibilities. For instance, when asked how to improve different problem situations, youth informants commonly answered 'fix it for us'; 'build cheaper apartments'; 'give us subsidies'; 'create work opportunities by arranging theatres, musicals etc.'; and so on. These responses illustrate a reliance on others to create and organise possibilities. We pondered whether this might relate to the long history of having state and municipal authorities that provided for the people in this part of the country. However, the same attitude was not found among adults, such as the social workers who also participated in the study.

The last question dealt with the respondents' visions for themselves as well as for the municipality, and relates in our analyses to the credal dimension. This set of answers made plain the significant difference between the respondents' visions for the municipality and for themselves. The visions of the municipality tended to be almost unrealistic dreams: Arvidsjaur was seen as a conglomerate to which big companies are drawn and where culture was flourishing and with endless entertainment possibilities. Compared to this, the respondents' visions for themselves were much more modest and mundane. They hoped to have a job, a family, a house, and a car. While we leave the reasons for this difference to future studies, our findings illustrate a significant 'disconnect' between visions for themselves and for their municipality of which they are a part.

In the above analysis, we have illustrated the applicability of the Dooyeweerd's modal aspects to an enriched understanding of a situation. Furthermore, we have illustrated that the modal aspects are related. This means that one cannot refer to one dimension solely without considering the impact on the others. For instance, what is due to people and what is experienced as justice can only be understood in the context of other dimensions. In Dooyeweerd's terms you cannot claim justice without considering the impact on others and this consideration can only be appreciated by, e.g., relating to the social, economic, and ethical.

An additional reflection relates to the fact that the answers often could be characterised as having their leading function in the social aspect. This may be due to the fact that the questions focused attention on social interaction or the answers revealed interest for the possibility to socialise - we cannot differentiate these two possibilities. So to enrich the variety in the answers, we sought additional leading functions and therefore could provide a richer analysis in terms of modal aspects involved. This effort made us more aware of young people's perception of interaction between modal aspects in, for instance, education. Here the education itself was not the determining factor, but often the possibility to see friends during school hours and thus possibilities for fellowship and fun.

By structuring the SSM Rich Picture according to the framework of modal aspects, we noticed the dimensions especially important - which dominated the situation in this specific context - as well as strong interrelations to other aspects. Consequently, aspects given less attention, or no attention at all, became visible. This structure for organising real-world situations was robust and highly suitable for analytical purposes, while it also maintained a holistic view. As shown above, different issues could be analysed, differentiated, and connected with the help of the modal framework. This facilitated an enriched understanding of the situation and permitted dialogue about different stakeholders' view. The significance of this finding is perhaps best illustrated by the reaction of one of the youth social workers involved in the study. He could see things in our rich picture that had escaped not only him but also the rest of the research team. The thing that astonished him most was that youth seemed to want authority but not the responsibility, nor even social integration. This episode supports Avison, Shah and Golder's statement that "rich picture diagrams are an excellent communication tool" (1993:313). By incorporating modal aspects and leading function, we have added further value to the process of appreciating the situation. Dooyeweerdian thinking was helpful in clarifying and advancing understanding of stakeholders' points of view and therefore optimized the learning potential and provided valuable insight for both the individual and the group.

5.   ENRICHING SSM'S DESIGN AND COMPARING PHASE

In this section, we continue the exploration of applying Dooyeweerdian thinking to Soft Systems Methodology techniques and in doing so we turn our attention to the phase of designing and evaluating Human Activity

Systems. In this technique, the focus is on modelling purposeful activity explicated in stakeholders' perspectives on the problem situation. The models include monitor and controlling systems in which SSM criteria for evaluating the process are supplemented with Dooyeweerd's modal aspects. For a fuller account of this SSM enrichment, see Bergvall-Kåreborn and Grahn (1996b). We apply the evaluation measures in a different setting compared to the previous section to illustrate the applicability of this approach to a variety of situations. First we begin, though, by showing the relation between SSM's original measures, and other features of the SSM process, to Dooyeweerd's modal aspects.

SSM's evaluation measures of Human Activity Systems resemble parts of the modal framework. In this chapter we explicate these connections in relation to the normative dimensions of Dooyeweerd's structure. See Table 1. The more determinative aspects are attended to in the hard systems methodologies and since the purpose is to explore the interaction to Soft Systems Methodology, we do not include them here.

Table 1. The modal aspects related to SSM measures of performance and process

MODAL ASPECT	SSM MEASURES AND CONCEPTS
Credal	E3 - effectiveness
Ethical	E4 - ethics
Aesthetic	E5 - elegance
Economic	E2 - efficiency
Social	Analysis Two (social aspects)
Lingual	Root Definition
Historical	The problem situation itself
Logical	E1 - efficacy

The first column in Table 1 expresses the normative aspects while the second lists SSM evaluation criteria, E1 to E5 and also some profound concepts within the methodology. As a result of this comparison between SSM measures and concepts and the modal aspects, areas of compatibility emerge (Mirijamdotter, 1998). The credal aspect links to E3, effectiveness, through the chosen *Weltanschauung* for the system ('W' of CATWOE). Effectiveness relates to the system's outcome in relation to W. The ethical aspect is found in the measure of performance, E4, ethics. The aesthetic dimension corresponds to SSM's measure E5. The economic refers to resource use and corresponds therefore to efficiency, SSM's E2. The social aspect represents the dialogue intrinsic in SSM, as guided by respect, consideration, and conventions. This dimension also explicitly considers SSM's cultural analysis of social aspects, termed Analysis Two. The lingual dimension with its kernel symbolic representation permeates the whole process. In SSM specifics, the formulation of root definition is an example of symbolic representation being clearly visible. The historical aspect predominates the design through its formative power, i.e. today's problems comes from yesterday's 'solutions', and is also a prerequisite for the future. Finally, the logical dimension coincides with measure E1 - efficacy. The concept of logic and the concept of efficacy are, in everyday language, usually perceived as each other's presupposition. If one is logical, one is also considered to be efficacious and if one acts efficaciously, the actions are presumed to follow a logical flow. As a consequence, if one wants to make a task more efficacious, this is usually done by looking at the present flow of activities that make up this task and then this flow is evaluated, in terms of logics, to find the best solution.

Table 1 illustrates the compatibility between SSM and Dooyeweerdian thinking while Table 2, below, demonstrates how Dooyeweerd's modal aspects can supplement and enrich SSM's original measures of performance - or, as we prefer to call them, performance indicators (PI), because 'measure' has too quantitative a connotation and 'indicator' better expresses the spirit of the discussions in the SSM comparison phase; indeed this term is already used in the SSM working glossary (Checkland, 1981:115) and covers both the qualitative and quantitative aspects in a more comprehensive way.

Table 2. Definition of the performance indicators to be used in SSM's modelling phase

PERFORMANCE INDICATOR	CLARIFYING QUESTION
Credal	Is the right thing being done? *
Ethical	Is the transformation morally correct? *
Juridical	Is the transformation just?
Aesthetic	Is it aesthetically satisfying? *
Economic	Is resource use minimum? *
Social	Is the social need accounted for?
Lingual	Is the communicative aspect considered?
Historical	Can we derive lessons from history about similar transformations?
Logical	Does the means work? *
	(*Source: Checkland, Forbes & Martin, 1990)

Thus, Table 2 represents the performance indicators for designing and evaluating Human Activity Systems when supplemented with modal aspects; they correspond to the modal aspects themselves, as shown in the first column.

The second column gives questions for discussion and reflection on each performance indicator and for defining appropriate evaluation criteria to judge and appreciate the model. The approach, illustrated in Table 2, offers a richer assessment of the issue under study, as we next illustrate with a real world example, the Estonia ferry catastrophe. The examples and illustrations given are based on information by the press and may include mere speculations and flaws. Since our aim is to illustrate the performance indicators, not to investigate the Estonia accident per se, this example intends to be illustrative only.

The Estonia catastrophe occurred during the night of 28th September 1994, when the ferry sank into the Baltic Sea on its way from Tallinn to Stockholm, carrying over 1000 people. Only 137 people survived. About 900 died in what is considered to be the worst shipping catastrophe in Europe since the Second World War. This calamity has left no one unaffected and the recurrent question is how something like this could happen. In the following, we use SSM to develop a conceptual model of an activity system for safe ferry transportation as depicted in Fig. 3. Then we discuss suggestions for improved performance indicators, drawing from our extended approach.

The figure shows the root definition and CATWOE-test for obtaining safe ferry transportation. We also include the conceptual activity system (activities 1-6) and its monitor and controlling sub-system (activities 7-9). The purpose of the system is to ensure safe ferry transportation in order to reduce emergency situations in contemporary shipping business. In accordance with the focus of this chapter, we concentrate on the measures used in the monitor and control activities. To clarify the contribution of our model, we discuss each performance indicator in relation to Fig. 3. We start by looking more closely at the logical performance indicator and proceed upwards in accordance with our model, Table 2.

In evaluating the logic of this system, the question 'does the means work?' (Table 2) considers whether the activities distinguished in the system produce the desired output. In order to reflect on this question, we must analyse our conceptual system by distinguishing the different activities and their order and, finally, check that the activities taken together can be counted as safe ferry transportation. Different levels of safety must be distinguished, because ensuring that the chairs are not damaged and that stairs and floors are not slippery do not alone guarantee safe ferry transportation. Early in the Estonia example, some investigations experts attributed the accident to the huge amounts of water that entered the car deck. This theory developed because, before Estonia left the harbour in Tallinn, the ferry was inspected by two Swedish Maritime supervisors in Tallinn for in-service training. They made a minor remark to the press about the inadequacy of the inner door seals but also said, when interviewed, that this alone could not have caused the disaster. Other experts thought that the hydraulic system that regulated the locking of the bow doors had not been in order; something the supervisors did not check. These observations imply that the Maritime supervisors did not investigate the logical relationship between different factors, i.e., what caused the inadequacy of the inner door seals. Reflecting on the casual relation might have led them to control the hydraulic system.

Root Definition:

> A system, shipping-company owned and staff operated, that runs on existing ferries and in existing weather, in order to achieve safe ferry transportation and thus reduce emergency.

CATWOE:
C - passenger, staff
A - staff
T - need for safe ferry transportation -> that need met
W - reduce emergency situations
O - shipping company
E - existing ferries; weather

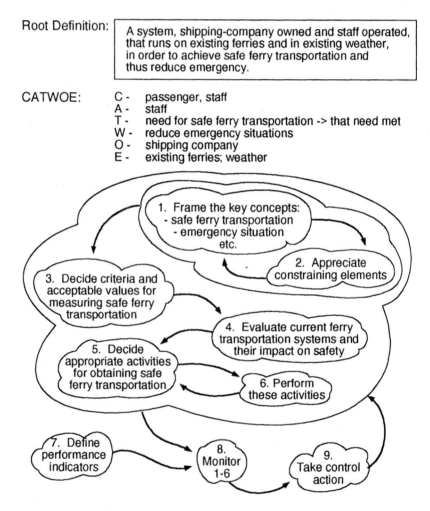

Figure 3. A SSM-representation of a system to achieve safe ferry transportation

When regarding the historical impact in the model in Fig. 3, the chosen question is 'can we derive lessons from history about similar transformations?' By studying events that have occurred in the past, lessons can be learnt. More specifically, the impact of the historical dimension can be seen in ferry construction. The use today of 'ro-ro' ferries (roll-on roll-off) is a case in point. Previously the construction was U-shaped with both entrance and exit in the stern. This has been replaced by the more efficient ro-ro design where the cars enter the stern and then exit through the stem. The design necessary for this type of economy means that they are vulnerable to filling up and then capsizing. The British ferry 'Herald of Free Enterprise' which capsized outside Zeebrugge in 1987 is another example of this potential problem. History, in the form of statistics, will also tell us that ro-ro ferries are more vulnerable then their precursors, so vulnerable that the Royal Institute of Naval Architects states that they are 'unacceptably sensitive' to damages. Also, according to Lloyd's Register of Shipping, statistically there will be one severe accident involving ro-ro ferries every fifth to seventh year.

The lingual aspect mainly concerns feedback information, understood in its full context, as well as data collection and generation. In regard to the lingual aspect in our example, questions could be raised such as: Which are the appropriate information systems? What information is needed? How should that information be channelled and communicated? Are key concepts, criteria, and decisions clearly stated and clearly communicated? In reference to Estonia, were decisions taken about safety routines clearly communicated to the staff? Were incidents reported back to decision takers? Were monitor and control systems appropriate and active? There were speculations, for instance, that the information systems monitoring the vehicle deck were turned off for the night so they would not disturb the staff. If this was so, discovery of water entering the car deck would have been delayed, precipitating delay in taking appropriate actions as well as in alarming staff, passengers and emergency centres.

In regard to the social indicator, our conceptual model requires the possibility for social interchange, necessity for social competence, and consideration of established roles. More specifically, when designing the actions which comprise activity 6, strategies like the following need to be discussed. Should operations be performed manually or computerised? Should they involve several people or should specific individuals be tasked? Should worker competence emphasize diversity or specialisation? Should tasks be attached to a bigger whole or carried out in isolation? The choice of strategy will affect the social aspect in the work design.

The economical performance indicator concerns the use of resources in relation to outcomes and can lead to discussions such as the following: Which constraining elements should be considered? How many current ferry transportation systems should be evaluated? Which are the key concepts that need to be framed? How many levels of monitor and control systems are needed? For each perspective there is a point when it is no longer economic to 'frame' yet another 'concept', to 'evaluate' one more 'ferry transportation system', and so on. This situation occurs when the costs are higher than the value it generates. In relation to Estonia, the owners cut costs by hiring cheap staff that did not have sufficient education, and by this they also decreased the value of the outcome. Because of this there have been speculations that the shipping company traded security for economical reasons. In hindsight, if this is true, it might have been more economic for the company to hire more qualified staff and by that perhaps avoid the catastrophe from happening or at least reduce its severity.

In regard to the aesthetic dimension, SSM assesses the elegance of the system by asking if the transformation is aesthetically satisfying. To illustrate the importance of the aesthetic dimension, we emphasise its kernel harmony. This can be done by consciously designing a system that harmonises the staff's working hours and free time. Separating working hours and free time is especially important on long travels where the ferry is the environment for both work and pleasure. Improving staff accommodations, rest and relaxation facilities, and socializing opportunities, and through that harmony between work and pleasure, could also improve crew readiness.

When evaluating the juridical aspect, the focus is on how the process and the output are affected by the laws of justice. Sea traffic is regulated in national laws and orders which are based on international agreements such as the International Safety Management Code initiated by the United Nation's International Maritime Organisation. In theory, most nations have homogeneous rules for sea traffic. What differs between them is the degree to which these rules are observed. Different shipping companies give varying attention to these regulations, and different countries have dissimilar sanctions for ensuring that the laws are obeyed. In activity 5 in our example, when deciding appropriate activities for obtaining safe ferry transportation, these regulations must be taken into consideration.

The juridical situation can be very complicated, as in the case of the Estonia, where three different countries, Sweden, Estonia and Cyprus, were involved. The ferry traffic between Estonia and Sweden was run by

Estline Ltd. which was registered in Sweden. At the time of the accident, Estline Ltd. was owned both by Estonian Shipping, an Estonian state company, and the Swedish shipping company Nordström & Tuhlin. Estonian Shipping and Nordström & Tuhlin also owned the company Estline Marine Ltd. which was registered in Cyprus. This company owned the Estonia and leased the ferry, without crew, to Estonian Shipping. Estonian Shipping, in turn, leased the ferry, now with crew, to Estline Ltd. To open the shipping route between Tallinn and Stockholm, Estline had to buy a ferry and therefore needed a bank loan. However, Estline wanted an Estonian crew and therefore planned to register the ferry in Estonia, but the bank involved required that the future owner of the ferry was registered in a western country. Therefore, Estline Marine Ltd. was created and registered in Cyprus. This company applied for the loan and bought the ferry, and the ferry was registered in both Cyprus and Estonia. The advantage to the company of registering in Cyprus was that, though being a western country, Cyprus, unlike most western countries, did not object to the ferry being registered in several nations, and did not demand the ferry be staffed with its countrymen. Consequently, in the case of the Estonia, laws and interpretations on sea traffic from three different nations must be considered.

The ethical dimension raises a multitude of interesting issues for discussion. The Estonia catastrophe provides many examples of the consideration of man for his brother even to the point of self-sacrificing action. When one of the shipwrecked was lifted by a winch from the raft to a rescue helicopter, he was so exhausted that he fell through the harness. One man from the rescue crew then jumped from the helicopter into the water and managed to get hold of the man despite the huge waves. He swam back to the raft with the man and managed to save both the man and himself. One would think that those who were saved, especially under such dramatic circumstances, felt lucky, but many of the shipwrecked people that survived felt extremely guilty. They asked themselves how they could be happy in such a tragic situation, and if they had survived at the expense of others, and if they could have done more to save their friends and fellow passengers. In our activity system, when identifying acceptable criteria and values for safe ferry transportation, ethical considerations play a big role.

Finally, to evaluate the credal aspect in our example the following needs to be considered. First, by asking whether our longer term goals are reached or not, the T (transformation) is checked in relation to W (*Weltanschauung*). More specifically, will meeting the need for obtaining safe ferry transportation lead to reduced emergency situations? By asking whether to strive for our longer term goals, our beliefs and assumptions about reality are critically examined. These reflections then generate new questions that further challenge our beliefs and assumptions. For instance, reducing emergency situations may be important for the shipping company since disasters cost money and create chaos. How does this compare to the importance of saving lives, which has an intrinsic value of its own?

As the preceding indicates, when inquiring into the W of the model, different underlying assumptions and beliefs on higher recursive levels can be identified, which makes that particular W meaningful. For example, the lasting effects of the Estonia catastrophe demonstrate the importance of conceptions in individuals' life decisions. After this calamity, many people lost their faith in ferries as a transportation means and were afraid to travel by sea. Consequently, the shipping companies in and around Sweden lost 30% of their passengers. In response, shipping companies went to extreme means to restore consumers' confidence. They welded the bow doors on the ferry while claiming that this was not really necessary since the construction of bow doors was safe. They even gave free tickets away to increase the number of passengers and thereby restore business traffic. The underlying assumption throughout was that as more and more people started to travel by sea again this would restore faith among both travellers and in society in general.

In Fig. 4, based on Fig. 3, we provide succinct questions for evaluating the performance of the system, for the purpose of summarising our thinking on how to design a system for safe ferry transportation.

The clarifying questions for each performance indicator above pertain to our example of a system to obtain safe ferry transportation. The questions are ranged in accordance with our model (Table 2) presented previously in this section. Benefits achieved by complementing SSM with further evaluation aspects illustrated that important aspects such as lessons from history, social implications and juridical connotations, were easily lost when the focus was limited to only three measures of performance, the 3 E's, suggesting how the system could fail. By considering additional dimensions, as was done above, we reveal unpredictable disadvantages, as well as unpredictable advantages. Through this greater insight into future notional situations emerged.

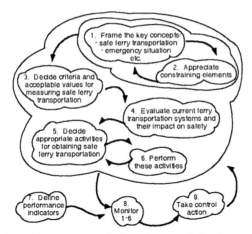

Log - does this count as safe ferry transportation?
Hist - what can be learned from one's own and other's experience?
Inf. - are concepts and decisions lucidly stated and communicated?
Soc. - are social needs respected?
Ec. - are minimum resources used?
Aes. - is aesthetics considered?
Jur. - how is the system affected by laws of justice?
Eth. - how do ethical considerations influence the systems?
Cre. - are the longer term aims reached and ought they be striven for?

Figure 4. Performance indicators defined for the system to obtain safe ferry transportation

6. CONCLUSION

Work to date suggests the usefulness of enriching Soft Systems Methodology with Dooyeweerdian concepts and frameworks to more fully capture multifaceted dimensions of problem situations. This has been demonstrated in two cases of which one was an application to structure data to get a Rich Picture of situation concerning youth in the inland of the North of Sweden. The second case was applied to the Estonia ferry accident and concerned a conceptual model for designing safe ferry transportation and performance indicators which both point to the design itself but also to comparing perceptions about design.

Both of these cases illustrate the richness and depth to which the inclusion of modal aspects to SSM techniques contributes. The key is that the modal framework is intuitively understandable to people, which encourage 'sensemaking', but still provides a richness through its completeness and interrelations, that allows for deeper inquiry. Simultaneously, intuitive understanding is also a quality of SSM, the foundation on which we build our expanded method. We conclude that incorporating the Dooyeweerdian framework optimizes the learning potential of SSM and enriches the understanding of stakeholders' points of view. By incorporating modal aspects and leading function, we have added further value to the process of appreciating the situation.

Since the nature of our thinking both shapes and limits the analysis and design, a variety of perspectives need to be included in the systems we design and evaluate. There is a mutual relationship between our designs and our worldviews. Consequently there is a need for conceptual frameworks that can manage the diversity and complexity of our design visions. For instance, benefits achieved by complementing SSM with further evaluation aspects enlarges our scope of inquiry and results in a clarified appreciation of relevant issues. Finally, as illustrated in this chapter, this approach has proven relevant in many kinds of organizations as well as, more generally, for assessing situations of interest.

To date, usage of SSM in interaction with Dooyeweerdian thinking has mostly been restricted to academic environments, in teaching and research. So far, it has not been integrated into organizational practice and learning and this is where our work is now directed. We currently explore the efficacy of teaching participants in problem situations to apply SSM thinking in combination with Dooyeweerdian thinking, which we term Multimodal Soft Systems Methodology.

The work includes developing teaching models which make tacit components of systems thinking sufficiently explicit and clear for adoption by not-formally-educated individuals and the teams of which they are a part. Our hypothesis is that when sufficient members of a problem situation can use systems thinking, the organization will self generate learning as, propelled by the exhilaration of successful problem solving, people organize themselves to design, implement, and sustain better systems within richer workplace environments.

Part II:
Socio-technical
Systems

Chapter 6

THE SYSTEMS CHARACTER OF MODERN TECHNOLOGY

Sytse Strijbos

1. INTRODUCTION

Technology normally calls to mind all the machines and technical appliances we surround ourselves with and use for a wide range of purposes. We turn on the television to find out what is going on in the world - locally and far away. We push our shopping cart filled with groceries to the cashier who, to the tune of little beeps, reads the bar codes on the items we have placed on the conveyor belt. With the mouse on our personal computer we click on 'print', and the article we have just completed rolls from our printer. With a pile of paper in our hands we stroll to the photocopier and run off a number of copies. We pick up the telephone to have a moment's contact with someone nearby or far away. With the click of a button we send our e-mails via the Internet and download the information we want to have. We step into a car, train or airplane to go to work or on vacation. In every facet of life, technology helps us to achieve our everyday objectives.

This description of our wonderful world of technology could be called naïve - not in the pejorative sense of the term, to be sure, but in the sense of what comes directly to mind from our everyday experience. Copernicus notwithstanding, we still say the sun comes up in the morning and goes down at night, for example. And this depiction has lost none of its meaningfulness for our everyday lives. Yet to understand certain phenomena we do have to know that the sun does not orbit the earth and that it forms the fixed centre point of the solar system.

Now, something of the same sort, it seems to me, applies to the way we speak about technology. Thus in order to gain a good grasp of the significance of technology in modern society, I shall in this chapter first endeavour to move away from the naïve depiction of technology expressed above and to penetrate instead into 'the real world of technology' that underlies our everyday experience in contemporary society (section 2). Then I will point out that the typical character of modern technology arises from its scientific foundation (section 3). Because this book focuses on computer and information technology I will consider next the question whether these newer technologies mean a fundamental change between the present structures of the technological society and those of the so called 'information society' (section 4). Finally, I will argue in section 5 that the systems character of technology which as such is a given of the modern age does not determine the evolution of our society. Since the rise of modern scientific technology we don't have the choice anymore for or against a world-wide technological society. However, we have still the choice to shape its structures for the benefit of a humane and cultured society.

2.   FROM TOOL TO ENVIRONMENT

Striking in the description of technology with which we started is the emphasis on material artifacts. It is a depiction focused on appliances, instruments and machines. Directly connected with this emphasis is a second characteristic of the naïve conception of technology. We use the apparatuses at our disposal to perform certain operations. We do things with them. Technology considered at the level of the artifact concerns the acts of individuals. The world of technology is portrayed as a collection of separate things or tools that an individual can utilize at will in order to achieve a certain goal. Just as we use a hammer to drive a nail into the wall or a rake and shovel to trim the garden in our spare time, so also, according to the naïve conception of technology, do we send an e-mail by computer and drive to work by car. The auto and the computer are nothing more than objects of human use.

It should not be surprising that the naïve conception of technology, which refers to the individually acting person and to everyday experience in the use of technology, originally permeated the philosophy of technology. Ernst Kapp (1808-1896), who may be regarded as one of the first philosophers of technology and the person who first used the term philosophy of technology, defended in his *Grundlinien einer Philosophie der Technik* (1877) the model of technology as the projection of human organs. To Kapp there is an intrinsic connection between the organs of the human body and the technological tools that people have produced in history. In his tools, man constantly produces himself. Thus the hand is the natural instrument that is continued in the diversity of hand tools bequeathed to us by traditional cultures. "The bent finger becomes a hook, the hollow of the hand a bowl; in the sword, spear, oar, shovel, rake, plow and spade one observes sundry positions of arm, hand, and fingers, the adaptation of which to hunting, fishing, gardening, and field tools is readily apparent." (Kapp [1877], 1978:45). Continuing in this vein with examples from our own age, one could say that the crane is a projection of the human arm and the bulldozer of the human hand.

What has been called here the naïve conception of technology fits traditional technology well. However, it does not enable us to grasp fully the phenomenon of technology in the modern world. Many who have thought deeply about modern technology have therefore abandoned the naïve view. Ursula Franklin published a fascinating book about "the real world of technology." In it she discusses the various models that underlie our ideas and our discussions about technology. "Technology," according to Franklin (1990:12), "is not the sum of the artefacts, of the wheels and the gears, of the rails and electronic transmitters. Technology is a *system*. It entails far more than its individual material components. Technology involves organization, procedures, symbols, new words, equations, and, most of all, a mindset."

In depicting technology as a system, Franklin represents a nowadays widely shared view. The systems thinker and biologist Von Bertalanffy observed, for example, that the very development of technology led to a different approach to technology, in terms not of independent machines, as was long customary, but in terms of systems. The auto, train and airplane, for example, depend for their use on an infrastructure or system. The auto presupposes a road network with a complex system of rules for its use, prescriptions for maintenance, financial regulations, and the like. On both the user's and the designer's side the systems character of technology makes itself felt. One makes use of the car, train or airplane together with others and in so doing participates in a collective traffic system. And viewed from the design side, one can say with Von Bertalanffy (1968:2) that "air or even automobile traffic is not just a matter of the number of vehicles in operation, but is a system to be planned or arranged."

The systems character of technology accordingly means recognizing the connections between material artifacts. The traffic system as a technological system is composed of a great variety of components. Thinking about technology in terms of systems also means seeing that such systems function through the participation of many different agents. The 'real world of technology' that underlies the many material artifacts surrounding us in our habitat is composed of countless systems. Technology is not a gadget or apparatus that stands separated from us as an external object, it is not a tool in our hands but it is the environment or 'the house' in which we all dwell today. Technology determines the public space of our existence. If we utilize technology by driving our car to work, for example, we participate in one of the systems within which modern society happens. And the same holds for a computer, a television, a pill, and our kitchen equipment. "Someone who flicks a switch is not using a tool. They are plugging into a combine of running systems. Between the use of simple techniques and the use of modern equipment lies the reorganization of a whole society." (Sachs, 1999:14)

That modern technology evinces different characteristics from traditional technology does not have to mean, of course, that conceptualization in terms of artifacts has lost its value altogether. Such conceptualizing should be called not incorrect but insufficient. For in modern technology the issue is *more* than just that of the individual relationship that people have with technical objects for the realization of certain ends. And the question that calls for attention is that of where that 'more' comes from. What makes modern technology so different? Or to put it still differently: what accounts for the signalized systems character of modern technology, referred to by the French philosopher of technology Ellul as one of the first, and because of which, according to him, we can typify modern society as a 'technological society'?

To seek what is specifically 'technological' of our contemporary society in separate things is not really satisfying, as just explained. Furthermore, it is difficult to deny that also in ancient times and in traditional societies people in their existence depended upon technological things they made themselves, including the cultivation of land, for example. It is thus not the material artifacts of modern technology that give our age 'technological' stature, at least not without more being said. The 'technological' must have to do with special characteristics that are missing in classical technology that upon closer examination we can discover in the things with which we surround ourselves.

Take for example an automobile. While it affords us a great measure of mobility, the cap on the gas tank reminds us at the same time that the freedom gained exists in dependency. The automobile can only fulfill its intended function if among other things we have access to enough fuel. And the moment petrol stations can no longer supply fuel, we all come to a standstill together. To extend the example of the automobile, this functions within a system of roads, gas stations, sign posts, but also for example of fiscal regulations, toll booths and much more. And the cap on the gas tank reminds us of that encompassing system on which the automobile depends in its functioning. One has to notice, however, that the systems character of modern technology is also inscribed into the artifacts themselves. The cap refers to an encompassing system on which the automobile depends but at the same time to the automobile as a system that itself forms a collection of parts that only work as desired if these are all correctly assembled. In short, modern technological artifacts are more or less complex material systems that are in turn interwoven into the complex patterns of a systemic technological environment.

The shift from 'artifact' to 'environment' since the rise of modern technology is related to its systemic character. Accompanying this shift, one may now add, is a far-reaching change both in the artifact itself and in the relationship between people and artifacts. The material artifact of modern technology is a whole constructed of components, a system that in its own turn functions as a subsystem in a broader, humanly constructed context or environment. And insofar as the individual person to artifact relationship is concerned, people too, just like artifacts, are no longer self-sufficient, independent entities but are inseparably connected to collective behavioral dimensions within socio-technological systems. Now, that brings us to the 'integrative vision for technology' and the broader concept of technology that is presented graphically in Fig. 1 of chapter 1. And again, the question that arises to confront us is that of what invisible factor X could have changed the character of modern technology and society so profoundly.

3. THE SCIENTIFIC FOUNDATION OF MODERN TECHNOLOGY AND SOCIETY

To answer the last question I want to refer to the ideas of the Dutch philosopher-engineer Hendrik van Riessen (1911-2000) and the French philosopher-sociologist Jacques Ellul (1912-1994), contemporaries both of whom developed a broad view of technology and society. As explained in chapter 1, an effort is made in this book to develop an interdisciplinary and integrative approach to technology by making use of elements from the systems sciences and some basic notions of Dooyeweerdian philosophy. Well then, Van Riessen developed his ideas in the Dooyeweerdian tradition of thought, while Ellul for his study of the phenomenon of technology used concepts featured in the emerging systems sciences of his day. In the continuation I will discuss differences and similarities between the thinking of Van Riessen and Ellul about technology with a view to arriving at a fruitful synthesis. In doing so I will concentrate heavily on the fundamental importance of science for technology in its modern form. I begin with a number of observations about Van Riessen's philosophy of technology.

Technology is not something like applied science, as so many still incorrectly think. It is true, to be sure, that scientific knowledge is applied in the development of technology: for example, calculations of strength are needed for designing a bridge. Even more fundamental than that, however, is the transfer of scientific method to technology. Technology, as Van Riessen (1949, 1979a) defined it, is the forming of inanimate material with the help of tools for human ends. In modern technology such forming is based on a scientific design. The designer who is confronted with a technological problem seeks by working in a scientific way to find a solution for it. In other words, the technological problem is transformed through analysis and abstraction into a series of elementary component problems. The solution of these component problems then leads to a series of elementary building blocks that the designer can use in varying configurations. The scientific method that is introduced into the designing is subsequently projected into the forming via the design. It is the way for forming. In technology the control of the whole, the material artifact, is realized through control of the parts. Distinctive of the scientific-technological method is a *neutralizing analysis or separation of functions followed by a varying or variety-producing synthesis or integration of functions*.

By way of illustration one may refer to the cap of the gas tank mentioned above as an elementary part of the automobile that is a result of a neutralizing division of functions, while the automobile is a final result of a variety-producing integration of functions. Summarizing in terms of systems theory one can say that the result of scientific-technological forming is a whole or a system featuring a hierarchical construction in which the parts from a lower systems level are integrated at a higher systems level. It is worth noting in passing that Simon (1996, chapter 8) in his essay about the architecture of complex systems that has become well known in the literature of systems science arrives at the same conclusion as Van Riessen, albeit by an entirely different line of argumentation.

While Van Riessen approaches technology at the level of artifacts and adopts a 'narrow concept of technology', Ellul defines technology more broadly (cf. Fig. 1 in chapter 1). Directly in the first pages of his book *The Technological Society* (1964) he warns the reader that by 'technique' we ordinarily mean, incorrectly, machines. The machine is according to him only a particular manifestation of technology. Technology in industrial production did begin, surely enough, with machines, but it now embraces much more and has in fact taken over the direction of all man's activity, and "not just his productive activity." "The term technique, as I use it" says Ellul (1964:xxv), "does not mean machines, technology, or this or that procedure for attaining an end. In our technological society, *technique is the totality of methods rationally arrived at and having absolute efficiency ... in every field of human activity.*"

Although the ideas of Ellul and Van Riessen are quite different, they stand in closer proximity to each other than one might think at first glance. Upon closer examination it turns out that in his definition of technique Ellul takes two modern powers together, technology and organization, that Van Riessen wants emphatically to separate. We have already seen that for Van Riessen scientific method is the foundation of modern technology. Well now, the same is true for him of modern organization. Characteristic for both these powers in their modern form for Van Riessen (1979b) is that scientific method lies at the foundation of each. And it is this common foundation that forms the bridge between his narrower and Ellul's broader conception of technique. That is clear from the last citation, in which Ellul in fact construes technique as method, and then, nota bene, as scientific method, which makes its initial entrance in the sphere of technological-industrial production. The scientific method (or more precisely: the scientific-technological method) that originates here goes on eventually to control everything: the economy, the social sphere, the psychic sphere, and so forth. And through the domination of the scientific method in all of society's facets, technology grows into an all-embracing power that threatens human freedom and seems to coerce people constantly to adjust to the demands of the technological system (cf. Ellul, 1980a).

Apart from the differences of definition of technology, the views of Ellul and Van Riessen about technology and society run parallel. In speaking of the technological society, Ellul means a society that is entirely in thrall to what in Van Riessen forms the foundation of both modern technology and modern organization. And while Van Riessen begins his philosophy of technology with an analysis of engineering practice and points out from that angle of approach the systems character of modern technological artifacts, Ellul focuses on the broader society and puts his full emphasis on the systems character of the social-cultural context or environment in which these artifacts fulfill their function. In both Van Riessen and Ellul, the scientific method is the invisible factor X we were seeking. Since the Industrial Revolution, the turn from science to practice has profoundly altered the character of technology and society. In the last century and a half, an enormous development has taken place, first in Europe and then worldwide, that carries society in the direction of an all-encompassing and global technological system.

It can come as no surprise that after the work of classical philosophers of technology like Van Riessen and Ellul round about the middle of the last century, the technological society and its further development have remained the subjects of a stream of studies. I have in mind in particular the publications of the nineties by postmodern sociologists such as Giddens, Beck and Baumann and the reflexive modernization and globalization and rise of the risk society that they discuss. It is to be anticipated that these and other themes will be found not stand apart on their own in isolation but that they may be understood better through consideration of their connection with the distinctive character of modern scientific technology. Thus I have shown, for example, how the dominance of modern technology and the process of evolving uniformity that accompanies its development could go hand in hand with a growing pluralization of society (Strijbos, 1997). And elsewhere I have argued that the same processes underlying the technological integration of the world and the globalization of society also seem to awaken new modes of global citizenship (Strijbos, 2001).

Further consideration of these latest discussions of today's society fall beyond the scope of this chapter. In the next two chapters Jan van der Stoep will take a critical look at the views of several postmodern thinkers concerning the importance of communications technology for society. With the central theme of this chapter in mind, I do however want to devote some attention in the next section to the debate of the seventies about the First and Second Industrial Revolutions and the rise of the Information Society (cf. Lyon, 1988), a debate that recently received new impulses from Castells (1996) standard work on the Network Society; namely, I want to examine the question concerning whether information technology means a fundamental break with the systems character of the modern world. Indeed, are we entering an entirely different kind of society as is proclaimed by trend-setting spokesmen such as Wiener, Bell and Toffler?

4. THE ROLE OF COMPUTER AND INFORMATION TECHNOLOGY

It is a widely held view that fundamental progress in machine technology, namely, the invention of the steam engine (James Watt), was decisive for the birth of industrial society during the so-called (First) Industrial Revolution. The Industrial Revolution is regarded in that case as a technological revolution. In his classical writings on industrial society Marx observed, however, that the first factories of the Industrial Revolution and the system of capitalism that went with them did not depend on the steam engine at all. According to Marx, the rise of industrial society is conditioned by structural changes in the organization of human labour that take place in a long process of development via different stages. Decisive for the Industrial Revolution is an invention in the organization of work, which Ellul (1964:22) later calls 'economic technique'. It is the technical work in the factory that brings modern machine technology and modern organizational technique together for the first time (cf. Pacey, 2000:18-19).

While modern organizational technique has its roots in the domain of economic production, it is also successfully applied in the later development of the industrial society, to many other sectors of society. The economist and systems thinker Kenneth Boulding (1953) was one of the first who draw attention to the fact that since the forties of the previous century organizational technology has received strong new impulses due to the appearance of the computer. In an early study he noticed that we were on the verge of a new organizational revolution of tremendous scope brought about by what he then called the electronic calculator. The invention of the computer with all its far-reaching implications for the technique of organization was also a source of inspiration for the American sociologist Daniel Bell, who gained renown especially for his book *The Coming of Post-Industrial Society* (1973). A typical new feature of the post-industrial society according to Bell is the creation of a new 'intellectual technology'. In the second half of the previous century, parallel with the rise of computer technology, there appear new scientific disciplines such as information theory, cybernetics, game theory, and systems analysis. Together these form the new intellectual technology that has made it possible to gain insight into the interaction of larger numbers of coherent variables and to get a grasp on problems of 'organized complexity' (Weaver):

> "....the methodological promise of the second half of the twentieth century is the management of organized complexity (the complexity of large organizations and systems, the complexity of theory with a large number of variables), the identification and implementation of strategies for rational choice in games against nature and games between persons, and the development of a new intellectual technology which, by the end of the century, may be as salient in human affairs as machine technology has been for the past century and a half" (Bell, 1973:28).

By speaking about a new intellectual technology, Bell is using the term technology here in a sense similar to Ellul's definition, namely as "the use of scientific knowledge to specify ways of doing things in a reproducible manner." From this standpoint the organization of a hospital, for example, can be regarded as a social technology. Well then, from the last part of the passage cited it is clear that for Bell a central characteristic of post-industrial society lies in such social technologies, in the control of "human affairs."

This significance of the new intellectual technology for the organization of social life remains a central theme in Bell, also in his later publications about the post-industrial society as an information society (Bell, 1979). What in essence he wanted to indicate by it he made clear by explaining that for centuries our consciousness of social reality was determined by our relationship with nature. Later, in the period of industrial society that unfolded during the last two centuries, our consciousness of reality came to be determined by our relationship with technology, which is to say with the things we use to control nature. Characteristic of the post-industrial environment, however, is the fact that reality is determined almost entirely by interactions between people. Thus first man interposed modern technology (machines) between himself and nature (industrial society). He thereby gained enormous power to control nature. Subsequently, however, with the rise of post-industrial society, social reality too became an object of technological control. As a result, social life has been absorbed more and more into large organizational structures established and regulated by man. In short, life transpires within organizations.

There can be no doubt that in addition to modern (machine) technology, which is based on science, modern organizational power or social technology, too, which is also based on science, is increasingly characteristic of contemporary society. One must acknowledge that Bell is correct in this regard. It is also correct that the previously mentioned cluster of new scientific disciplines, born in the second half of our century, offers fundamentally new possibilities; compare, for example, computer simulations of models in which the interaction of many linked variables can be investigated and the natural scientific experiment in which it is only possible to investigate the relation between two variables. But this does not yet demonstrate the correctness of Bell's concept of post-industrial society. For it is decidedly incorrect to maintain that modern organizational power would be a typical phenomenon of the present age of systems thinking and computer technology, a phenomenon that arose thanks to the possibilities of a new intellectual technology. Not only the scientific-technological control of nature but also the organization of social life with the help of science and technology is, at least in principle, already present from the earliest beginnings of the Industrial Revolution of the eighteenth century. In fact, the invention of organizational technique was decisive for the (First) Industrial Revolution, as we have already noted.

From the foregoing one can conclude that Bell's new intellectual technology has not created a society with entirely new structures. Between industrial society and Bell's concept of post-industrial society there are gradual differences. Of course, one must not ignore the fact that there are impressive changes in society. But what can now attain development with the help of computers and information technology was already present in a simple or embryonic form from the beginning of the (First) Industrial Revolution. Therefore one can speak of a continuation, expansion and intensification of the technological society with which we have been familiar for quite some time now. Post-war developments in the field of the systems sciences and information technology - Wiener's (1954:136) Second Industrial Revolution - have created new possibilities for an enormous expansion and for the perfection of the organizational-technological control of society. Or, as Bell (1973:33) puts it:

"The goal of the new intellectual technology is, neither more nor less, to realize a social alchemist's dream: the dream of 'ordering' the mass society. In this society today, millions of persons daily make billions of decisions about what to buy, how many children to have, whom to vote for, what job to take, and the like. Any single choice may be as unpredictable as the quantum atom responding erratically to the measuring instrument, yet the aggregate patterns could be charted as neatly as the geometer triangulates the height and the horizon."

The question arises to what extent Bell's dream of ordering our mass society has already been realized. With Ellul one must admit that all aspects of life in society have been drawn into the sphere of technology. With the characterization of our society as a 'technological society' Ellul has in mind a contrast with the 'industrial society'. He agrees with Bell that the latter type of society which grew out of the (First) Industrial Revolution now lies behind us. To be sure, one can maintain that the "present-day society is still industrial; but," so Ellul (1980a:2) says, that is not its essence." The technological changes with which we are now confronted are of considerably greater importance for the structure of society than ever was the case in the past. Ellul therefore criticizes Bell and also Touraine for being insufficiently aware that technology is the essential new phenomenon of our time. In industrial society, modern scientific technology still had only a limited place. In our own day and age, that is no longer so. According to Ellul, technology in the technological society is no longer comprised of separate elements but due to the computer has assumed the character of a system, of a purposefully organized whole (Ellul, 1980a, chapter 4).

Although there are substantial differences between Ellul and Bell, both authors thus stress the overwhelming importance of organizational technology and of computer and information technology for the structure of contemporary society. Ellul too holds that there has been a turn in the development of industrial society. His view of the turn from industrial to technological society is already evident from his discussion of the phenomenon of technology in his study of 1964. This matter is raised again and elaborated further in a sequel of 1980 entitled *The Technological System*. It is *"the computer that allows the technological system to definitively establish itself as a system. ... For example, the urban system can close itself up only because of the urban data banks. ... Computers are the correlation factor in the technological system."* (Ellul, 1980a:98,101) Does Ellul mean to say, in speaking of the technological system, that one can identify society with a technological system? Is society transformed by technology into an enormous 'megamachine' (Mumford, 1967, 1971)? Is Wiener right to construe society as a cybernetic system? No, Ellul (1980a:17) regards such views as not only dangerous but also obviously in conflict with the facts. "It is quite easy to prove with facts that our society is not mechanized. On the other hand, it is full of short circuits, jammings, chaos, and also huge nontechnicized voids; on the other hand, man in this society has not really been mechanized to the point of being just a gear."

Thus Ellul makes a strict distinction between society and the technological system. These two cannot simply coincide. Society, after all, is made up of real people, while the system is abstract. Therefore it could only be in an extreme situation that society and the technological system would coincide. And no one can seriously assert, so Ellul says, that we have already reached that point. The actual situation today is such that one must picture a society in which a technological system has been installed. There are numerous tensions and conflicts between society and the systems character of the modern world. The situation is rather like that of the problem of the environment. Just as machines can disturb the natural setting by causing various kinds of pollution, so the technological system can disturb the social environment. The result is then "disorders, irrationalities, incoherences in society." (Ellul, 1980a:18)

5.   FINAL NOTES

One must admit Ellul that Bell's dream about the ordering of the mass society with our new technologies has up till now not resulted in a harmonious world. Its shortcomings and problems can be interpreted, at least in part, as disturbances of social life caused by the systems installed in society. While social nature is pre-existent to the technological system, it seems that the growth of technology does not leave the social body intact (Ellul, 1980a:81). The question which can be raised with respect to Ellul's view on the situation of our society is that on his standpoint it is not clear whether there is a solution, a liberating perspective for the serious problems. Is it possible to reconcile the organizational-technical systems in which we have to live today with the nature of the human individual and our social nature? Can our new intellectual technology (Bell) help us to tackle the problems?

In a somewhat different vein as Ellul also Van Riessen expressed his concerns about the future of our society. While in Ellul's view modern technology (the scientific method) as such is the problem, Van Riessen opposed against the domination of the scientific-technological method. He was keenly aware of the limits and boundaries of the scientific method in technology and in scientific-technological control. These boundaries already appear on the side of inanimate matter. It is never entirely possible to eliminate the unicity of the materials that are used in technological forming. Certain variations always inevitably appear in the composition, however small these may be. Moreover, the circumstances in which the materials are processed are never perfectly constant because of disturbances in the surroundings, no matter how far the isolation of the process may be carried. The moment people appear in the technological forming, something else is added as well. One must then take into account typical human characteristics and also normative limits for the application of the scientific-technological method. The designer is then confronted with the question of how to take people's creative freedom and responsibility into account and how the work situation should be organized in technological forming.

While he saw the lurking danger that the boundaries and limits of the scientific-technological method may be ignored, Van Riessen (1957, chapter VI) warns against what he calls 'the abuse of planning' and 'the collectivization of society' in which people lose sight of the free and responsible person. However, just in connection with this central point his analysis in my judgment falls short, or at least Van Riessen's thought lands up in an otherwise justifiable critique of the absolutization of the scientific-technological method. He does not address the question concerning whether it is possible to develop a scientific method and methodology that are able to do justice after all to human responsibility and to the normative aspects involved in fitting out a new human habitat.

Here I believe is to be found one of the reasons for the fact that while Van Riessen has certainly bequeathed to us an original and interesting philosophy of technology, he has not worked concretely in the practice of technology with his philosophical insights, enriched though these undoubtedly are by his engineering experience, so that his philosophy evinces a certain lack of empirical content that has also not gone unnoticed by others (Kroesen, 1999). Yet why would it not be possible to develop a normative approach in which the problems of the social habitat that is stamped by technology can be fruitfully addressed in a scientific way?

This is precisely the point that this book desires and is challenged to answer. In other chapters in this volume I will discuss the implications of the gained understanding about the systems character of modern technology for an ethical approach of technology. In chapter 12 I will discuss the idea of a systems ethics guiding human acting in the shaping of our technological world. Such a systems ethics becomes concrete in the different systems methodologies that have been developed in the past decades as will be discussed in chapter 14.

Chapter 7

COMMUNICATION WITHOUT BOUNDS?

Jan van der Stoep

1. INTRODUCTION

One of the major socio-technical systems in our late modern, or postmodern age is the communication network. This communication network forms the backbone or infrastructure of our society and delivers the instruments to create a public space in which people converse and associate with each other. Its main goal is to support social interaction by lingual means, thereby connecting people over huge distances in time and space. The communication network does not only include technical artifacts like televisions, computers, data networks, telephone lines, satellites etc., but also consists of an important organisational component: institutional arrangements, management techniques, informal networks etc.

This chapter develops the thesis that the design of the communication network is at least partly the product of a worldview in which people are seen as free and independent individuals, constantly driven by a longing for unlimited accessibility. People become aware of their own situatedness; they try to make contact with other cultures, they develop new forms of social cooperation and they create their own way of life. This paper presents an ideal type, that is, a specific construction of history based on selected facts and a method of exaggeration. It calls attention to an important spiritual drive behind the development of our network society that is worth considering. The Internet, as we will see, fits incredibly well the patterns and demands of our high modern society. It may be said to embody postmodern thought and to bring it down to earth (Turkle, 1995:18). Yet I want to avoid simple causal explanations, as if the communication network is the materialised product of a certain ideology or worldview. Although worldview is an important determining factor in the development of new technologies, it is not the only one.

When, for example, the US defence staff developed the ARPANET, they were not able to forecast that the same technology would become the new backbone system of a global society. There is no reason to deny that the development of the Internet has its own logic, technologically induced rather than socially determined, providing and commanding new forms of social interaction (Castells, 1996:9-61).

Sections 2 to 4 explore how the ideal of unlimited accessibility unfolds in modern western society and how the drive to expand one's scope of action is supported and accelerated by the development of new technological means. I distinguish three historical periods that are usually associated with the subdivision of modern history into an early modern, an enlightened modern or industrial modern and a late modern or postmodern phase, each characterized by its own technological devices, the map and the clock (early modern), the factory and the conveyor belt (enlightened modern), and the mobile phone and Internet (postmodern), respectively. At the end of the chapter, section 5 in particular seeks to demonstrate that the view of man as a free moving individual in time and space is on the one hand an important spiritual drive to emancipate people from traditional social hierarchies and cultural boundaries but that, on the other hand, it also may lead to a form of hypercommunication. When one does not acknowledge the situated and embodied nature of human agency and the intrinsic normativity of communication practices, the newly gained power to organise human social relationships in a more individualistic and flexible way, may become counterproductive and self-defeating. The reader is warned beforehand that this paper does not provide a list of recommendations about how to improve our current situation. That is not the purpose of this philosophical endeavour. This paper offers an alternative, more realistic way of regarding human communication, thereby clearing the ground for an analysis of the normative structure of the communication network in chapter 8.

2.   EARLY MODERN NAVIGATIONAL TECHNOLOGIES

In order to understand the rise and development of the global communication network we have to go back to the advent of modernity in the sixteenth and seventeenth century and explore how early modern people envisaged space and time and developed new technologies of navigation in order to have access to the ends of the earth. Early modern society was characterised by a new way of picturing the world that culminated in the development of the modern map and the modern clock.

In traditional cultures, time and space are envisaged as structured according to a given social, cultural and religious world order (Van der Stoep and Kee, 1997:400-401). Each point in time is regarded as having a given meaning related to the moment of the day, the succession of the seasons and the course of the liturgical year. Men and women living in traditional societies are obliged to follow the prescribed rhythms and rituals of seed time and harvest, of being born, maturing and dying. They have a collective and cyclical sense of time. They do not plan the future in a calculated way but follow the rhythm of the seasons and simply take care for what is to come by following the traditional customs and fulfilling their daily duties (Bourdieu, 1977:19ff.). With the introduction of the modern mechanical clock, however, all moments are seen as equal points on the same linear time-scale. The mechanical clock and the modern calendar, provide an abstract and neutral standard for measuring time. Time is seen as something which has to be 'filled in', an empty space of moments which has to be given meaning by human activity. This abstraction of linear time from the actual moments of the day can never be realised in full; nonetheless, its impact becomes widespread as a result of the standardisation of calendars, including the imposition of the western Christian calendar on a world-wide scale, and the standardisation of time across regions: Greenwich mean time, global time zones, winter and daylight saving time.

The same is true for the traditional sense of place. Characteristic for traditional space is its subdivision into centre and periphery, high and low, sacred and profane, dark and light, inside and outside, hot and cold etc. Men and women move carefully within a culturally and religiously conditioned environment, knowing that each place has its own meaning and that people all have their own place in a pre-structured order (Bourdieu, 1980:441-461). This traditional sense of place is reflected in premodern maps, which were drawn from a concrete point of view to depict the centre and periphery, sacred and profane places etc. In the modern experience of place, on the other hand, all places are seen as essentially equal to one another. People are free to move within the world, not hindered by cultural or religious meanings and prohibitions (although even in modern society national boundaries and private properties set a limit to the freedom of movement). In the modern map, the world is presented as an abstract system of coordinates in which the different local places are interchangeable with each other, only varying in their degrees of latitude and longitude. Just as in the modern experience of time, so too in the modern experience of place, the world is seen as an empty space to be filled in and given meaning by the endeavours of man (Giddens, 1990:18).

The modern experience of time and place does not stand on its own but is strongly related to the disembodying of the human ego. The human person is seen as a point in space and time, free to explore the surrounding world. One is not (or at least one thinks one is not) hindered any longer by social, cultural or religious rhythms or boundaries. One early proponent of the modern humanistic self-awareness, René Descartes, stated that it is better to build a city or a governmental constitution at once according to a rational and methodical plan than it is to elaborate upon a traditional patchwork of houses and streets and upon the insights of earlier civilisations (Descartes, 1983:93-95). By the use of a scientific technological method (technique in the sense of Ellul; see also chapter 6) Descartes abstracts from complex reality and subdivides every problem into a series of elementary components, trying to find a solution by making a synthesis between the different decontextualised building blocks. Another early proponent of the modern disembodied ego is Michel de Montaigne. For De Montaigne travelling is a way to encounter other cultures, critically examine one's own tradition and culture and write one's autobiography. Not the rationally planned environment but the individual collection of experiences is his central focus point; he thereby abstracts the individual traveller from his situatedness in time and space while rejecting every sense of a pre-given world order.

The early modern map and mechanical clock, together with additional instruments like the compass, were artifacts of a new, typically western form of navigation. In traditional South Pacific navigational practice, all sense of motion and direction is measured from an embodied perspective in which one refers to actual, existing objects and circumstances; early modern Europeans mentally stepped out of their own situatedness, saw the world as something in front of them, and located where they were from a disembodied, overhead perspective, a god's-eye point of view (Ihde, 2002:6-7). By informing us about time and space, modern navigational technologies enable us to orient ourselves not only in the local environment we are familiar with, but also in foreign areas we do not know yet, thereby facilitating voyages of discovery, intercultural encounters and trade on a world-wide scale. However, although with the new technologies of navigation and exploration we might have the impression that we are travelling in empty space, we are still embodied and situated people. That is already manifest in the structure of the modern clock and map. In the design of the mechanical clock, the numbers on the dial and the position of the hands still refer to the turning of the earth around its axis. The same is true of the design of the modern map. One is only able to orientate oneself in space with the help of a map if the map refers to actual, existing spatial objects like rivers, roads, borders and mountains.

3. MODERN INDUSTRIAL AND SOCIAL ORGANISATIONAL TECHNOLOGIES

The next step towards the realisation of a modern communication infrastructure coincides with the age of the Enlightenment and the development of new technologies to organise social relationships in time and space. The society of the Enlightenment is an industrial society characterized by a huge differentiation of social functions. The standard example is of course the factory. With the application of scientific technological method not only to physical nature but now also to social reality, namely to the organisation of workers, the production process is subdivided into a series of elementary actions, and all these elementary actions are then integrated again into a fixed temporal-spatial framework: the structure of the conveyor belt. This raises the level of production enormously, and thereby also the standard of living. With the division of labour, work time and free time, work place and place of residence are also strongly demarcated from each other. Labourers are disembedded from the local situation and context in which they make a living and are reintegrated into the abstract organisation of the modern factory, where they are characterised primarily by their specific position and function. The rise of industrial organisation goes hand in hand with an enormous increase in scale. Human activities are structured not on a regional but on a national level. The nation state as a political entity becomes more and more important. With the nation state, a national market and a national school system also develop. The nation becomes a large organisational whole with uniform standards, money equivalents, a national curriculum etc. Areas and regions are differentiated by function, and a strong demarcation is made between cities and rural areas.

An important milestone in the rise and development of the nation state was the French Revolution, which saw the bourgeoisie seize sovereignty and overturn the traditional government of King and Church. The citizen as *l'homme universel*, speaking the official language of the state, became the new norm. In modern Enlightenment thinking, political identity is based first of all not upon the identities of the different ethnic communities living within the national border but upon the active participation of citizens in the process of democratic decision making (Habermas, 1994:636). Instead of being determined by birth, social position is gained by the citizen's own effort and competence. The human person is not seen as part of a family, community, or ethnic group in the first instance but rather as a free individual who rationally plans his or her own career. One is responsible for one's own life and is not a mere extension of one's family or ethnic group. The gain of personal freedom did not, however, prevent new forms of social stratification from developing that were based not on birth but on profession and social class. The French Revolution, for example, coincided with the imposition of the Parisian dialect as the official language on other regions in France, making the Paris bourgeoisie the role model for modern national citizenship (Bourdieu, 1991:46-49). The challenge of modernity, therefore, is to develop a cosmopolitan spirit in which universal values like freedom, equality and human dignity are practiced beyond the boundaries of the local community without introducing, in a more hidden way, new forms of social exclusion.

The division of labour and the nationally organised differentiation of social functions calls for new forms and technologies of social organisation. An important example is the mass transportation delivered by the national railways. Using timetables, the railway company plans in an abstract and rational way where trains arrive at fixed moments (Giddens, 1990:19-20). Then there are the means of mass communication, which integrate people not in a physical but in a mental way. One cannot underestimate the role of the press, radio and television in the development of modern national consciousness. Nowadays, with the use of satellites and the global hegemony of broadcasting companies like the BBC, CNN and MTV, mass communication has even been an indispensable factor in the development of a global cultural awareness, generating for the very first time in history a world-wide public space. Even on an individual level, people are provided with new instruments to organise their social lives, instruments that facilitate the integration of different social roles within each person's own project of life. With the help of diaries, they make appointments and arrange them rationally in time and space. Of high importance too is the national telephone network, which connects the different places in which people dwell, facilitating communication over long distances. Before visiting someone at home or at work, people nowadays usually first make a phone call or send an e-mail message.

Modern industrial society is a highly differentiated society in which human activities are scattered in time and space and in which individuals have to travel over large distances. Control over time and space means control over one's own life: liberated from local bounds, individuals enjoy an enlarged radius of action. Modern people experience the automobile as a symbol of their newly gained freedom (Sachs, 1990). It gives them a virtual unlimited freedom of movement to visit interesting places and events and to explore natural reserves and other areas that lie beyond the reach of modern planning; it even makes living outside the city possible. Yet however important this newly gained freedom is for organising one's life and enlarging one's scope of action, one must not forget that the ideal of the free-floating modern individual living his or her own life as a free project is simply an abstraction from ordinary life, in which all kinds of people do have all kinds of different actual experiences of time (Bourdieu, 2000:224-227). The rationally organised time of the manager or the free time of the modern tourist differs enormously from the stressed working time of employees coping with deadlines, dates and timetables or the dead time of the unemployed living their lives as a 'wasted' period. The sense of space of the modern jet traveller flying over the Indonesian archipelago is totally different from the sense of space of the traditional mortals down below, where people living in their small villages and fields are locked up within the bounds of their traditional culture (Berger, 1979:1-3). The ideal of a disembodied ego planning its own life in time and space and demanding unlimited and immediate access to all kind of places is based on an elitist illusion that excludes a large group of people and that imposes a typically western way of life upon people with other cultural backgrounds. It is not without reason that in automobile commercials the driver almost always travels through an empty space, not hindered by fellow drivers, intersections or other physical obstacles. Pictures of stressed out fellow drivers, traffic jams, busy city centres or filled parking lots remind us too much of ordinary life and so have no place in the modern dream of unlimited freedom and accessibility.

4. LATE MODERN TECHNOLOGIES OF SELF-ORGANISATION

The so-called late modern or postmodern situation is characterized by the blurring of boundaries once set by industrial society: work/home, public/private etc. Instead of moving through a static matrix of social functions, the late modern or postmodern individual creates his or her own environment. postmodern space and time is flexible space and time. Where the modern politics of emancipation was meant to liberate people from the imperatives of tradition and religion and the illegitimate domination of other groups, postmodern life politics has become instead a politics of lifestyles and life-planning (Giddens, 1991:214-217). It is a politics not first of all about justice, equality and participation but of individual choice and self-actualisation. The self becomes a reflexive project. It is not seen as a fixed identity or a part of a larger whole, but as an identity under construction, a subject using the technologies of contraception, psychological therapy and plastic surgery to construct its own life narrative or biography. And instead of seeing itself as member of a group, organisation or class, the individual self builds its own relationships, its own social context. Even the most deeply held convictions of life become a matter of choice, of picking and choosing elements from a variety of religions and worldviews. In order to describe this new postmodern condition, some thinkers use the metaphors of the vagabond and the tourist (Bauman, 1993:240ff.). According to these thinkers, both groups of people journey through an unstructured space. They do not stay for long at a certain place, but structure the site they happen to occupy at a certain moment and dismantle the structure again as they leave. The only difference is that the tourist, in glaring contrast with the vagabond, lives his extra-territoriality as a privilege and as a way of distinguishing himself from others, and not as a form of exclusion.

Another way to understand our postmodern condition is to apprehend that space is annihilated: chronopolitics supersedes geopolitics (Virilio, 1998:17ff.). The individual self does not position itself within a pre-structured space, but constructs its own biography, its own life story by blurring the boundaries between different spheres and connecting them again in an interactive way. Ironically the same development may also be understood as an annihilation of time in which past, present and future are compressed into a single moment. That means that the different contexts through which one moves are simultaneously present, so that one easily switches between them at one and the same moment. It is especially the new computer mediated forms of communication that make such a form of flexibility and self-organisation possible. One may start by thinking first of the interactivity achieved by remote control and the computer mouse. With the remote control of the television set in hand, one easily switches between different channels: broadcasting is superseded by narrowcasting, and mass communication by targeted group communication. The same kind of interactivity is found on the world wide web: with a single mouse click one easily hops from one site to another, picking one's own way through the text. Secondly, one may go on to think of the interactive accessibility that is supported nowadays by the fast development and implementation of mobile phones. Before the invention of the mobile phone, communication was bound to fixed local points in space. In order to speak to someone by telephone, one had to call them, depending on the moment, at home or at the office. Now, however, people are directly connected to each other by mobile phone. Being connected to the global communication network is more important than being at the right place at the right time. A third and last example of interactive accessibility is provided by e-mail, newsgroups and chat boxes. These means allow large groups of people to be reached simultaneously. They also allow people to send and receive messages wherever and whenever they prefer. People no longer have to be physically present at a certain place and time in order to communicate with each other.

Where the technologies of modern social organisation once liberated individuals from the hierarchic structures of traditional society, the late modern technologies of self-organisation liberate them from the static grid of industrial space and time, thereby blurring all kinds of boundaries and distinctions. People may now reach each other anywhere and anytime and so build their own relationships and identities. Paradigmatic, and may be also symptomatic, for such a postmodern construction of social identities are the Internet nerds who communicate and play with each other in multi-user domains or multi-user games. In each of a number of windows on their computer screens they build a different identity and then easily switch from one window or identity to another. For such nerds, real life becomes just one more window, and not usually the best one (Turkle, 1996:13). The postmodern egocentric world, however, is not only a world of virtual construction and arbitrariness, it is also a harsh and risky environment in which one has to construct one's own biography, willingly or not, and in which every failure and misfortune is seen as a personal failure (Beck, 1992:136). And, what is even more important and dangerous, the Internet allows marginalized groups, nowadays especially young Muslims, to build their own networks of inferiority, in which they may construct their own framework of reference, distributing fundamentalist and extremist ideas on a world wide scale without control of official institutions and traditional hierarchies (Roy, 2002:165-183). Fortunately the actual horizons of human experience can never be wiped out fully. Just as the early modern clock and map still referred to actually existing places and moments, the flexible configuration of cyberspace still is bound to actually existing cities and metropolises as centres of innovation, communication and culture (Castells, 2002:224-231). We must not make the mistake of assuming that people in our late modern age are really persons without context and history. The self as a reflexive project, for example, is still situated, as a typical representative of a class of academics and free individuals exempted from the exigencies of everyday life (Bourdieu, 1986). Such a late modern self does not embody a neutral and independent ideal; rather, it presupposes a given moral framework within which free individual development is understood as meaningful and valuable (Taylor, 1991:38-39).

5.   COMMUNICATION AND HYPERCOMMUNICATION

In the preceding sections we have seen that in response to a typical modern longing for unlimited accessibility, that began in the age of exploration and discovery, people have developed all kinds of technologies to expand their scope of action and to liberate themselves from the narrow bounds of traditional society. They have constructed extensive socio-technical systems in order to communicate with each other across the boundaries of their local situation and culture. This has really enriched the quality of human life. People are not meant to stay in a certain place, like flowers rooted in the soil, but they have legs to move around, to meet each other and discover the treasures of the earth. Due to our enchanted relationship with technology, however, we late modern people often tend to project ourselves into the artificial worlds we construct with our technologies and then seem to forget that, just as the coordinate system of the modern map is an abstraction from the actual, existing places in which people dwell, so too both the rationally constructed grid of industrial society and postmodern cyberspace are mere abstractions. Our socio-technological systems of communication certainly change our human practices and self-understanding, but that does not mean they also change our situated and embodied human condition. Cyberspace might become just another machine fantasy in which we project ourselves as if we were part of the abstract reality of our technologies. Therefore, we must not ignore the fact that those technologies are but extensions of the human body that do no more than disclose reality in a specific way, thereby supporting us in our everyday lives (Idhe, 2002:85-87).

An important consequence of an overly one-sided concentration on the techno-organisational foundation of modern society (I use the terminology of the introductory chapter) is that we do not pay enough attention to the meaning of our activities. We often know very well 'how' to do something in the most effective and efficient way, but we do not have an answer to the question 'why' we are doing it (Winner, 1990:60-62). We have at our disposal in the means of communication a wealth of possibilities to reach other persons, but they only function in a useful and reasonable way if we also have a sense of what we have to say to each other, a message important enough to communicate. Unlimited access does not necessarily mean better communication or improved social interaction. Consider the excessive use of mobile phones. While people use mobile phones to communicate more often and longer with each other, the information density of their calls seems to decrease. This fits incredibly well the law of inverse proportionality of communication, already noticed by the nineteenth-century philosopher Søren Kierkegaard (1978:64), who predicted that with the new means of transportation and communication our decisiveness and our capacity to communicate would decline: "Suppose that such an age has invented the swiftest means of transportation and communication, has unlimited combined financial resources: how ironic, that the velocity of the transport system and the speed of communication stand in an inverse relationship to the dilatoriness of irresolution." The new means of communication help us to break through the boundaries of traditional life, but without a sense of our own situatedness, this newly gained freedom to communicate with each other may easily lead to hypercommunication, that is, to an overdose of communication that reduces instead of improves the interaction between people (see for a similar argument, Van der Stoep and Kee, 1997).

First of all, hypercommunication may arise when unlimited accessibility leads to a situation in which people only communicate with their fellows and peers and are no longer able to transcend their own group and mental framework. The existence of global networks may indeed support interaction between cultures, but such networks are only a necessary, not a sufficient condition for intercultural exchange. They may even frustrate a dialogue between groups. Emigrants, for example, may use the new means of transportation and communication to maintain contact on a regular basis with their family and friends, which makes it more difficult for them to come loose from their own traditional background. And with the use of dish aerials Moroccans, Turks and Arabic people living in the United States or the European Union may still watch their own Arabic television stations, oriented towards their background culture and not the culture of their new homeland, thus generating their own subculture and sharpening cultural contrasts. Such a process of cultural fragmentation is even more probable and dangerous in a situation of global injustice, when, for example, one dominant culture unilaterally imposes the technological and institutional framework in which communication takes place. This seems in fact to be the case when the imposition on a global scale of American popular culture (MTV, Hollywood, Coca Cola) leads to the annihilation of traditional cultures and the development of a global monocultural landscape, which inevitably generates its own reactionary counter forces (Mander, 1996:350-353,357).

Secondly, hypercommunication may arise when the new means of communication penetrate too intrusively the private sphere and when the demand for openness and public accountability frustrates personal, informal interchanges between people. Not every kind of conversation or social relationship lends itself to being made public; people need a sphere of safety, a personal space in which they may trust each other. Even in public and economic life such a zone of intimacy and trust is necessary. The only way to solve political dilemmas and conflicts is to have a selected group of people negotiate behind closed doors where they can suggest compromises to each other without the immediate intervention of the television camera. And as every manager knows, the most creative and innovative ideas are born during informal exchanges between employees who meet each other in the corridor or around the coffee machine. Immediate accessibility may liberate people from the confines of their local situatedness, to be sure, but it also means that people are caught up into the constricting networks and surveillance systems of journalists, managers, fellow citizens, politicians, etc. Unlimited accessibility may lead, in other words, to unlimited control that destroys the conditions under which communication between people may flourish.

Thirdly, hypercommunication may arise when people constantly create new identities and relationships for themselves and in doing so lose contact with their own biography, with the real world in which they breathe and make a living. In seeking to be an authentic person, one must not be constantly re-inventing oneself. One needs a sense of belonging, a social context that provides a framework of reference, a constellation of moral values by which to orient oneself and give direction to one's life. When we only build identities and relationships without integrating them into a coherent narrative embedded in a common history with other people, we communicate ourselves to death. This is especially true for the

Internet nerds mentioned above, who live on the screen and forget their own social situatedness and so live a life without past and without future. However creative and inventive they may be, nobody can doubt that they lack an authentic self and that their longing for appreciation and recognition will never be stilled. Unlimited access to all kinds of virtual spaces hinders them from becoming integrated personalities, sound persons on whom people may rely. Virtual life, therefore, has to be seen as a mere extension of real life and not the other way around.

6.   CONCLUSIONS

The argument of this paper is not that the advent of hypercommunication is an inevitable historical development or that communication technology is counterproductive in the end. What I have said is that when we are driven only by an unqualified longing for unlimited accessibility, we lack a directional device and so are no longer able to draw the line between good and bad uses of the new communication technologies. We must moderate the flow of our communications in order to reserve time to invest in social relationships and the communities in which we participate. That does not mean we have to return to pre-industrial or even premodern society. On the contrary, our highly advanced socio-technical systems may also help us to attain the goal of a more genuinely communicative society. For example they may provide us with the means, such as voicemail, SMS services or the trill alarm on mobile phone sets, that we need not only to expand but also to regulate our sphere of communication, thereby generating new opportunities to create a zone of intimacy for the things that really matter in life as well as in human relationships. In chapter 8, therefore, we will explore how the new means of communication may serve to enhance social interaction between people. At the same time, we will develop an integrative and normative view of technology.

Chapter 8

NORMS OF COMMUNICATION AND THE RISE OF THE NETWORK SOCIETY

Jan van der Stoep

1.   INTRODUCTION

In chapter 7 we have seen that the communication network is not a neutral socio-technical system but the outcome of a historical development driven by the modern strife for unlimited accessibility. Now, in this chapter, I want to examine another normative dimension of the communication network, the fact that it has to meet norms of communication in order to function in a proper way. This means that the building of a network society has to be adapted to the standards of human communication instead of the other way around. If engineers, designers and policy makers really want to ameliorate human life, they need an integrative vision for technology, taking into consideration not only the technological but also the social consequences of their activities. The use of new communication networks may only become beneficial and salutary when those experts not only ask the question 'how' communication has to be organised effectively: the techno-organisational foundation; but also 'what' it means to communicate with each other in a good and reasonable way: the lingual and social qualification (see the introductory chapter for a further definition of an integrative vision for technology and the distinction between the founding and qualifying function).

Unfortunately, a lot of communication theories in standard books are developed from a reductionist point of view, explaining communication by analogy of the machine. This common understanding of communication finds its origin in the cybernetics of Norbert Wiener, who describes communication as the copying of an amount of information from one context to another. By using a medium of communication, for example language, telephone, or radio, the sender, according to those theories, transmits a message, a body of information, to a receiver. The information exchange is seen as successful when the transmission of information is complete and the output approaches the input. In the approach elaborated in this book, however, communication is seen as a human activity, with an intrinsic normative character. In our daily conversations, we are all aware of this normativity when we distinguish for example between a clear presentation and a fuzzy one, an interesting book and a boring one or a pleasant dialogue and a difficult one. When I talk about *intrinsic normativity* I do not want to say that the norms of communication are to be found in the things as such, somewhere beyond the scope of our human imagination (the realist perspective). It is more that I want to emphasize that we cannot talk with each other without acknowledging certain rules regulating our social interaction. People, for example, have to trust each other that they are telling the truth, that they do not violate each others integrity and are ready to pay attention to each others concerns. Without such a sense of normativity, it is even impossible to have a real dialogue (Habermas, 1994:34). This also becomes visible by the fact that when we try to demonstrate that a lot of conversations indeed do violate the laws of truthfulness and trustworthiness, we cannot avoid to appeal to those values in order to make our point clear (Habermas, 1983:90-93).

The argument of this chapter is as follows. In section 2, I will examine how the new communication infrastructure provides new forms of interactive accessibility, thereby also changing the techno-organisational foundation of human communication activities. I will demonstrate that computer mediated communication in fact changes the writing and reading of texts as well as the patterns of social behaviour. Then, in the following two sections, I will bring in two normative perspectives as a counterbalance for a technicist and/or postmodernist view, in which the medium is seen as the message (McLuhan, 1964) and in which huge social transformations are forecasted because of the rise of information and communication technology. In section 3 I develop a hermeneutical perspective according to which words, phrases and texts can only be understood fully when we are familiar enough with the context in which the message is produced. We know first of all by acquaintance and only secondarily by formal utterances or statements. In this way I like to do justice to the fact that all language is characterised by meaning, and that communication is only successful if people understand each others concerns, goals and intentions. In section 4 I examine human communication from the perspective of critical social theory. According to this perspective communication is a way in which people create a public space, including as well as excluding certain forms of human behaviour. Critical social theory helps us to understand communication processes on a social level and makes clear that human communication only flourishes when people stick to their words, when they are honest, kind and respectful to each other and have social tact.

2. COMPUTER MEDIATED COMMUNICATION

As we have seen in chapter 7 a rather recent and highly influential change in the techno-organisational foundation of communication and social interaction is the rise of interactive accessibility, the possibility to construct a flexible network of information and communication, independent of fixed points in space and time. Interactivity in this sense does not stand for a kind of two-way traffic in which sender and receiver respond to each others messages. These kinds of interchange were already available on a large scale in the age of mass communication, facilitated by telephony and postal services. When we speak of interactive accessibility we point towards a kind of interaction, not between users, but between the user and the medium of communication. A good example is the use of the remote control in order to switch between programs and television stations, tuning in on the messages one likes to receive. Although even in the age of mass communication certain primitive forms of interactivity were already available and even necessary: the volume control on the radio, the dial of the telephone etc., nowadays, due to computer technology, the opportunities to interact with the transmitter are far more advanced and increase dramatically. This is especially due to the integration of information and communication technology. The advantage of computer technology is that, due to electronic switching, it may rapidly and faultlessly process a large amount of data and therefore may be used, not only as a thought instrument, but also as an instrument to control information and communication processes (Schuurman, 1980:19-23). With the help of computer mediated communication, we not only are able to shape the information content of the messages produced or received (subsection 2.1), but we also are able to shape the patterns of communication (subsection 2.2), integrating one-to-one (telephone), one-to-many (broadcasting) and many-to-one forms of transmission (inquiries, polls) in one and the same digital network structure (Van Dijk, 1997:12-16).

2.1   Electronic writing and reading

Lets start with the first way in which computer mediated communication changes patterns of human communication: the lingual practices of writing, reading and information storage. Due to computer mediated communication the writing process and the process of printing and reproducing texts changes drastically. Electronic writing or word processing may be seen as the next step in the succession from clay tablet, pencil and paper to type writer and printing press. Using software programs like Word Perfect or Word, authors are less bound to texts already written or typed down, but may constantly interact with their text, changing the text the way they want. They think on the screen, constantly arranging and rearranging texts, copying, cutting and pasting them again (Heim, 1993:5). But not only the production and reproduction of texts, also the reading of texts is transformed. One of the most important revolutions in this respect is the advent of hypertext. By the use of hypertext different subtexts in a database are linked to each other, so that one may easily jump from one locus to another by moving the cursor to a sign, a word or a picture and to click on it with the computer mouse. One browses through the text, following one's own line of thought. The use of hypertext, therefore, changes the static and linear structure of the printed book, with its fixed sequence of paragraphs, sections and chapters into a dynamic and non-linear network of simultaneously available subtexts (Turkle, 1995:17-18).

The impact of computer mediated communication on writing and reading is easily underestimated. Changes in the production of information, for example, have a huge impact on the relationship between author and publisher, as well as between the secretaries and their principals. The chain between thinking, writing and publishing becomes shorter, less typewriting is necessary and also individual persons now do have the possibility to publish an immaculately printed book or newsletter. Even more important, and to a certain extend may be comparable with the invention of the printing press in the 16th century, is the non-linear intertextuality of electronic texts, which gives the reader a more instrumentalist relationship with the text, reading the different subtexts in the sequence he want and having access to all texts and subtexts at one and the same time. The scientific method, underlying modern technology, with other words, also penetrates the activities of writing and reading, abstracting the different subtexts from the context of the original text and integrating them again in a new framework of understanding.

While the printed book abstracts the text from the actual context of oral communication, giving no reference to the voice or the original handwriting of the author and presenting the word as something written down before us, the hypertext also abstracts the reader from the fixed context of the written text, presenting the text as a substrate for knowledge production, as something to be manipulated and rearranged. To digest the text the reader does not have to get into the text, following the line of argument from beginning to end, but he can pick and choose between texts and construct his own message. In a certain sense this is not new. All people who process a lot of information like politicians, managers, scientists or other knowledge workers know that the only effective and efficient way to appropriate a large body of knowledge is by browsing through texts, jumping from one section to the other and taking the table of contents more as menu or entry to the text than as a means to find the place where one has stopped reading. The difference, however, is not only that such a way of picking and choosing is facilitated by technological means, but that the hypertext adds a new additional dimension to the text, detaching the reader from the original context of text, thereby prompting the reader to a more associative and a less hermeneutical way of reading (Heim, 1995:30-31). The electronic Bible, for example, only gives us entry by words, phrases, references etc. and does not present the text to us as a whole, as a series of books in which the story of creation, fall and redemption unfolds itself. This may enrich study and meditation, giving additional information as well as a better understanding of the text in relation to other texts, but it also may alienate the reader from the text because it stimulates a more subjectivist way of reading, actively searching one's own meaning instead of being kept in a more open and sensitive way by the expressiveness and eloquence of the text.

2.2   Interactive communication

The second way in which computer mediated communication changes the patterns of human communication is on the level of social interaction. Due to the development of an interactive network of communication it is easily to switch between one-to-one, one-to-many and many-to-one forms of transmission. Lets develop this insight a little bit further. First of all we have to acknowledge that not all kinds of interactivity are computer based in a strict sense. Mobile telephony, for example, is made possible by more advanced ways of data transmission, disconnecting the mobile phone sets from a fixed grid of local source stations, connecting them to repeatedly different stations when the user changes location. By use of this technology, mobile telephone sets, notebooks and laptop computers are integrated into a flexible hyperstructure adjusting itself to the movements and activities of the individual actors. Secondly, it is important to emphasize that the integration of the different forms of transmission is also accompanied by new forms of information exchange and social interaction. With the use of software programs like Netscape and Explorer a world wide web of advertisements, texts, and databases is generated in which all kind of data are accessible from every thinkable location on the earth. And Outlook, another software program, allows for the possibility to send and receive electronic messages to every thinkable person or group of persons, also making possible many-to-many communication (newsgroups). Thirdly, it is worth mentioning that with the digitalisation of data transmission new technical applications become available to regulate communication processes such as electronic address books, voice mail, calling number delivery and selective mail rejection. At the same time, this development also lead to new applications for regulating the accessibility of websites, databases and other media: pass words, paid access, firewalls etc.

Due to the new applications of computer mediated communication, social interaction is more and more abstracted from the local community and face-to-face conversation. This is a change not only in thought but also in practice, being supported by new technical devices. Letter, telephone and television already were means to communicate over large distances, thereby breaking through the spatial nearness of face-to-face conversation. Interactive communication, however, also bypasses the spatial and temporal configuration of social places and functions. It does not only blur the distinction between home and work as well as between private and public sphere, but also provides individual consumers with new means to build and regulate their own circuits of communication (chapter 7). Even the addition of audio-visual means and webcams does not compensate fully for this abstraction and decontextualisation of human communication, because as interfaces they not only make the other visible, but also deliver him with instruments to present himself in a highly sophisticated way, for example by manipulating the virtual representation of one's physical presence. In the same way as electronic writing and hypertext, the electronic means of social interaction encourages an instrumentalist attitude towards communication, isolating individuals from the social context in which they live and integrating them again into flexible patterns of social relationships. Instead of taking the social communities in which one dwells for granted, one purposefully build one's own network, determining one's own target group and bypassing all kinds of social hierarchies. This instrumentalist and functionalist way of communicating, however, does not by definition lack a personal or informal character. Especially in the personal sphere, nowadays, one is purposefully searching for pure relationships, true love and interesting (virtual) dates. And SMS, mobile telephony and e-mail often are used in the private sphere regulating rather personal and intimate contacts. What is at stake is thus not the shrinking of the private sphere as such, but the invasion of the private sphere by all kind of electronic media thereby easily giving personal relationships a more constructivist and arbitrary character.

3. THE MEANING OF INFORMATION

When we only envisage the techno-organisational foundation of human practices, as we have done in section 2, it seems that our communication activities change dramatically, replacing a static pattern of conversation by new forms of interactivity. In this section I discuss the presumed changes in behaviour from a normative perspective, especially focusing on the lingual dimension, thereby using an hermeneutical perspective. In the next section I continue the discussion by taking into consideration the social dimension. According to the postmodernist prophecies the increased accessibility to data calls for totally new forms of learning and knowledge transfer. Due to the proliferation of information, so the postmodernist prophecies hold, the individual self is freed from the restraints of tradition and ideology, composing its own personal opinion by picking and choosing between a variety of impressions, thoughts and idea's and integrating them again into its own framework of reference (Giddens, 1991:80-88). The dark side of this presumed development, however, is that people become disorientated because of information overload and because of a reduced awareness of the distinction between actual reality and virtual reality, between the real world and the world of knowledge construction. A lot of postmodernist prophets therefore warn us that we must not ignore the contrast between the virtual and the real world, using the first one only as a tool of imagination and not as a substitute of actual reality (Heim, 1995:135-136).

When we, however, look from a hermeneutical perspective, taking into consideration the lingual qualification of the communication network, we find out that stories, texts and documents are first of all means to structure the complex order of reality and to point towards those things and topics which are of significance and need more attention than other things. Such a perspective tempers the high expectations and fears surrounding knowledge production and point towards a direction in which the new means of communication may stimulate instead of frustrate the exchange of thoughts and ideas. An important advantage of the hermeneutical perspective is that it makes a distinction between having information about the world, i.e. knowing all facts and data, and knowing the world by experience, having insight into what is significant (Borgmann, 1999:14). Having information about the world has to do with knowing the world in an indirect way, knowing the world by description, that is by intermediation of signs or words ('savoir' or 'wissen'). Having insight into what is significant, however, implies that one also knows the world in a direct way, by acquaintance, without intermediation of signs, representations etc. ('connaître' or 'kennen'). We may also call this 'tacit knowledge', knowledge which is embodied, which has to do with learning by experience, feelings, intuitions etc. Having information about the world, although secondary to knowing the world by acquaintance, is an important means to disclose, broaden and transform our embodied and often locally oriented experience (Borgmann, 1999:1). Without maps, reports and records our experience about reality remains limited and closed. Information not only means a reduction of the complex order of reality, but also an understanding and disclosure of reality in a specific direction, thereby pointing towards the importance of things and the meaningfulness of reality. In order, however, to understand the meaning of formal signs, words or texts, they have to be re-embedded in actual practice. One has to interpret the information in relation to the context, the situation in which it was produced.

What does this hermeneutical distinction between knowledge by acquaintance and knowledge by description means for the world of Internet. Interestingly the world wide web, not only provides us with an overabundance of information, but also with new means to structure and integrate this information into a meaningful network of interrelated texts. Instead of a huge and unordered repository of arbitrary data, the world wide web must be seen as a network of more and less significant texts, related to each other by hyperlinks. This network of texts may function as a context of understanding, giving people a better insight into the importance and relevance of the different words and fragments, reintegrating the information provided into their own embodied and situated knowledge. Let's first examine the consequence of this thought for the information producers. Just as the author of a printed text, the

Internet publisher has to be aware of the central focus of his message, as well as to the way he has to order the different subtexts into an integrated and meaningful whole. Only when a clear and well organised scheme of thought and ideas is provided, the reader will be able to find his way through the hypertext, distinguishing the significant from the less significant, and having a clear notion of the structure of the document (for an similar argument related to interactive computer games, see Friedman (1995:85)). Besides that, when an author wants his site to be read, he has to link it to specific connection points on the Internet, for example Internet portals or other Internet sites which give a systematic entry to all kinds of relevant pages and which usually make a selection of the most interesting sites available. Simply publishing on the Internet doesn't have much result if one cannot rely on an institution that adds its reputation to your words or on a community of Internet publishers or other writers who are willing to refer to your text. One needs to make all kinds of associations in order to give an official legitimation to one's message and credits oneself with the authority to speak in the name of an institution, a group of people or a community of experts (Latour, 1987:103-144; Bourdieu, 1991:107-116).

The use of the Internet not only demands a strong discipline from the producers, but also from the readers of the texts. Instead of purposeless surfing from text to text and only following one's own interest, one must thoughtfully pay attention to the context in which the different texts are produced. In order to depend on what is said, one has to know the background of the author: what is his status? To which institution is he affiliated? Is his contribution recognised by other Internet users or not? One also has to have a meaningful framework of one's own, in order to put the information into perspective : What is the historical or scientific significance of the subject? How does it relate to other facts and interpretations of reality? How does it fit into one's own world view? Lastly, and not the least important, one has to have a sense of the stubbornness of reality, the fact that there is always a discrepancy between our thoughts and imagination and the reality in which we live. We learn the most as our horizon of experience conflicts with the horizon of others, and we have to search for a new perspective on the world, in which the former conflicting points of view are accommodated and a new fusion of horizons is realized (Taylor, 1995:149-151). Besides that, the most important and significant experiences one gains in situations in which reality does not always answer our expectations or desires. The virtual experience of climbing a mountain or encountering a tiger, how realistically produced, cannot compete with a real life situation in which one really is in danger (Graham, 1999:158-160). And instead of the sexual intercourse with a real person, the interaction with a virtual simulacrum is boring and addictive, because of the fact that every ground for being recognized, trusted or rejected by the other is missing (Coolen, 1997:55-58).

Browsing on the Internet and collecting a body of knowledge about a subject is more than only collecting data from various Internet sites and integrating them into a new whole. In order to select and evaluate information, distinguishing what is important and valuable, one has to have an insight into the actual situation in which one lives and has to know and understand what has to be done in such a situation, that means one has to have wisdom (Van Riessen, 1970:14). This presupposes a full grown power of discernment and sensitiveness which is only learned by acquaintance. Moreover it presupposes a love for truth and truthfulness, a very personal relationship and commitment with the object of knowledge, which is not gained in a short instance, but by a lifelong attention and concentration (Hart, 1984:227-229). How useful computer applications may be in the class room, more important is the personal enthusiasm and engagement of the teachers, who often function as real life examples for the students, encouraging them to enter into a certain profession and become committed practitioners. Of course, often in daily life, one has to browse through a large amount of texts, picking out those passages that are of interest. But when one is coming to the core of one's profession, or even more personal, to the core of one's own convictions, one cannot remain on the surface of the screen, hopping from text to text, but one has to enter into those texts, taking the time to understand them or even learn them by heart. Especially in our fast information age, in which we are prompted to formulate almost immediately an accurate judgement, while at the same time being confronted with an ever growing amount of information, practical intuition or 'tacit knowledge' becomes more important then ever, in politics, as well as in management, science and journalism.

4.   COMMUNICATION AND SOCIAL DIFFERENTIATION

We now turn to the issue of social interaction. As with information exchange and knowledge production, postmodernist thinkers predict that computer mediated communication will radically change our social contacts and relationships. In some way or another they all assume that the modern welfare state will be replaced by a network society in which relationships are build around free floating individual biographies and in which the personal lives of people are no longer integrated within a static social hierarchy. This, they hold, produces a rather insecure and risky social environment in which people enter and quit relationships whenever they want, and in which permanent positions are replaced by flexible jobs (Beck, 1992:127-137). At the same time, by the use of electronic media, society penetrates into the private sphere of individuals, who more and more are eager to share the most personal feelings and thoughts with their fellowmen, thereby making the intimate sphere transparent and turning the monitor or webcam into an instrument of expression and exhibition as well as an instrument of cybercontrol (Virilio, 1998:22-28,69-79).

What, also in this case, is lacking in the postmodernist utopias and dystopias is a sense of the intrinsic normativity of social interaction. People are envisaged as individuals, without socio-cultural background, strategically searching for the most interesting and promising connections and contacts. In most instances, however, the conversation partners share a common interest to keep the conversation going. In order to realise this, they have to be honest to each other, obey the rules of conduct ('netiquette') and show a willingness to invest in their mutual understanding. Greetings, utterances of respect and admiration as well as showings of hospitality are not, or at least should not be, empty phrases but forms of behaviour which remind us to the fact that we have to acknowledge the other as being important and valuable (Young, 2000:57-62). A basic willingness to keep in contact with each other and obey the rules of conduct is even a prerequisite for newsgroups and virtual communities in which people cannot hear or see each other directly, giving their contact a certain degree of anonymity (Bahm, 1995). Although in some Multi User Domains people are playing with identities, changing their character time and again, this is characteristic of only a small number of virtual groups. Beside that, the people participating in those groups all know and agree that they are playing with roles and identities and that they don't have to take each other too seriously.

It is an important insight in critical social theory that the interaction between people only flourishes when people function in a variety of different social relationships, so that they are not locked up in one of the different social spheres and society is not ruled only by the law of the market, the law of the state, or the law of the family or church. Already the 17th century philosopher Blaise Pascal proposed that society is organised into separate social spheres and that in order to prevent tyranny, the relative autonomy of the different social spheres has to be guaranteed (Pascal, 2000:79-80). Because every social sphere has its own form of censorship, its own specific way of distributing social authority and responsibility, a form of complex equality is maintained, in which the authority of one group of people, for example the scientists or the writers, balances the authority of the other group of people, for example the political leaders. When, however, the differentiation of social spheres is not respected and society is reduced to one type of community ties (the clan or the nation), or one type of inter-individual relationships (the market) the counter balance between the different social actors is destroyed and a selected group of people, having authority and monopoly in one sphere, may dominate the whole of society according to their own specific rule (Walzer, 1983:17-20). Of special importance for our study are the social-philosophical distinctions between public and private sphere and between the public and private good. I purposely use here the term public in two different meanings. In the first case the public sphere of the media is opposed to the private or intimate sphere of the house; the open space outside, instead of the closed room inside. In the second case the public is what is of public concern, what is in the interest of all people and not only of a certain person or group of people. The new means of computer mediated communication, however, seem to support first of all the invasion of the private sphere by the media, and secondly the dominance of the private law of the market over the public law of the state.

Lets start with the invasion of the private sphere. Supported by a drive to make everything transparent the private and intimate sphere more and more is penetrated by the media. Real life programs like Big Brother or The Osbournes on MTV, demonstrate that the media are indeed able to blur the distinction between the private and public in such a way that no distinction between public appearance and private personality is left. But instead of being the rule, those programs are more a deviation to the rule, transforming every extraordinary person into an ordinary man and calling for even more extraordinary television in order to compensate for the dullness of looking to the banality of real life. Just as people need inter-individual relationships in order to break through the strangleholds of the local community, they also need a sense of belonging, a family, group or region to which they belong and in which they feel at home. It does not, therefore, has to surprise us that globalisation and the development of international networks and organisations is accompanied by a new awareness of regional and national identity and the formation of so-called neo-tribes, groups of people actively distinguishing themselves by referring to their ethnic, national or cultural rootedness (Lash and Urry, 1994:314-326). Another example in which the search for maximal transparency is present in the behaviour of reporters and journalists who try to unveil every secret political negotiation and who do not respect any closed door. To a certain extend this is indeed their profession. Journalists have to fight corruption and the entanglement of interests, but when they drive this concern too far, they too much invade the relative autonomy of the political sphere, thereby not only turning every discussion behind closed doors into a conspiracy, but also devaluating their own profession into a searching for media hypes, making news instead of reporting, evaluating and interpreting the actual social events. One of the most important virtues of a good journalist, in fact, is that he knows exactly how to balance his professional curiosity with the prudence of a reporter, respecting the authority and integrity of the other professions.

Secondly, the new means of communication stimulate the dominance of private interests over the public good. By pointing towards the seemingly inevitable transformation of mass communication into interactive communication, postmodernist thinkers and marketers predict the end of the public domain. Public television will be replaced by Internet television, and broadcasting (focussing on the whole nation) by narrowcasting (focussing on specific target groups, all with their own life style). The laborious introduction of Internet television and pay-TV, however, seems to support another conclusion. Although media convergence is technologically possible and we may in fact receive television, radio and Internet in one and the same electronic device, their social functions are far too different to become really integrated. People use television for entertainment, the radio as a companion, Internet for their content-oriented information and they also like to read a classically printed book (Castells, 2002:193-194,199). Therefore we have to take into consideration the social setting in which television functions. First of all television is a passive medium, used by people to entertain themselves, to be witness of major social events (sport, national ceremonies, weddings and funerals of public idols and dignitaries) or to hear things which are of general interest. They not purposely search for information, like when they are behind the computer screen, but look to the television from a distance, evaluating and interpreting the things happening. Secondly the television functions as a public forum. It make people aware that they are part of a larger whole and it integrates them into a national or international community. Although media convergence and market forces point towards a privatisation of the public sphere, no modern society can do without public broadcasting services, providing people, from an official and more or less independent point of view, with information about the situation in the world. To know what is really going on, we needs a public space of news provision, not biased by private interest or contaminated by rumours.

5.   CONCLUSIONS

In this chapter I tried to demonstrate that computer mediated communication indeed thoroughly influences our activities of writing and reading as well as our patterns of social behaviour, but that does not mean that we, willingly or not, are entering a society in which all that was solid melts into the air. How new the rise of Internet and other forms of computer mediated technology may be, they are still subjected to lingual and social norms, and therefore are not beyond ethics (Lyon, 1997). The Internet is not a game without rules (Van der Stoep, 1998). The most acute danger for our society is not computer mediated communication as such, but the possibility that we let ourselves be let by postmodernist fantasies, which try to convince us that the medium is the message, and which let us oscillate between adjusting ourselves to our machines, or fighting against the unbeatable tyranny of the new logic of cyberspace. The normative framework developed here, opens another direction in the field of systems ethics, an ethics of transformation, searching for the intrinsic normativity and meaning of each social system and trying to transform each social system in such a way that it supports its function instead of frustrating it (chapter 12). Instead of an ethics of adaptation, in which we adapt our human communication to the machine, or an ethics of liberation in which we try to restrict the working of the machine as much as possible, we have to develop an Internet ethics in which, dealing with issues like intellectual property rights, criminal law and global justice, the main goal is to transform the communication network into a proper functioning infrastructure that is an integral part of our human activities and therefore in its structural configuration has to incorporate the standards of human communication.

Part III:
Human Practices

Chapter 9

EVALUATION OF SYSTEMS IN HUMAN PRACTICE

Albert E. Vlug and Johan van der Lei

1. INTRODUCTION

The occasion to this study is the wish to evaluate a national drug-safety system we build in the Dutch IPCI project. We want to evaluate this system in a way to be reproduced and criticized by others. We will start with a short description of the drug-safety system to be evaluated.

In the Netherlands, a General Practitioner (GP) plays a central role in the health care system. Patients usually visit their GP first and the GP usually starts a treatment, for example, prescription of drugs, referal to a specialist, or just arguing 'not to worry'. Patients receive prescribed drugs from the pharmacist, after medication control. Yet, side-effects may occur and patients, pharmacists as well as GPs may report these to a national institute. The analysis of these reports requires large databases and highly educated researchers in order to decide whether or not a drug is associated with the reported complaints. Several artifacts, like electronic databases, in the context of socio-technical systems, such as a national drug-safety system, support different human practices, for example, patient care. As shown in previous chapters, the environment of human practices has become systemic. That implies human practices are mostly interwoven with all kinds of systems. Evaluation of human practices encompasses therefore the evaluation of (socio-technical) systems.

A drug-safety-system belongs to the health system and inherits the ultimate goal of increasing the patients care. The specific goal of a drug-safety-system is to assess the safety of drugs. We distinguish between safety of drugs in development (the pre-marketing surveillance) and safety of drugs prescribed to patients (the post-marketing surveillance). The traditional method of the pre-marketing surveillance consists of clinical trials in which new drugs are tested on animals and young healthy volunteers. The traditional methods of the post-marketing surveillance are the case-control and cohort studies, in which patients using drugs 'in daily life' are observed. In the IPCI project we focus on the post-marketing surveillance (PMS). Post-marketing surveillance concerns the investigation of (side-)effects of drugs after the drugs have been introduced into the market. In the development of this system we distinguished between the information system on one hand and the organization of the context or society in which this system must operate on the other. Fig. 1 shows the information flow of the information system.

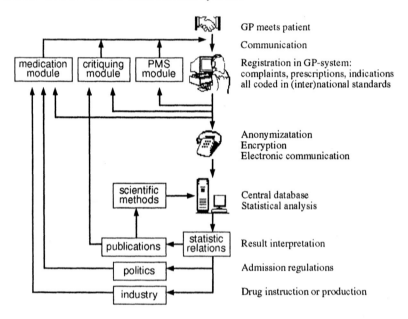

Figure 1. Information flow in the drug-safety system

The information flow starts with the patient consulting a general practitioner. During or after the consultation, the GP records some information in his GP-system. An additional PMS-module controls the recording adequacy of complaints, prescriptions and indications for prescriptions. Each month, the computerized patient records are sent to a central database. According to several statistical methods, many patient records enables a researcher to calculate and compare the (side-)effects of prescribed drugs. This research may result in publications. To close the circle, we presuppose a threefold feedback loop. Firstly, publications may result in critiquing modules that supports a GP to prescribe the most effective drug with the less side-effects under the specific patient conditions. Secondly, the government may take back the official registration of drugs with relative high risks and continue the admission of the most safe ones. Thirdly, the pharmaceutical industry may change or stop the marketing of drugs that appear to be legitimate suspicious. In the end we hope the therapy supported by this information system will benefit to the patient. Fig. 2 shows a simplified organization scheme of the social context in which the information systems operates.

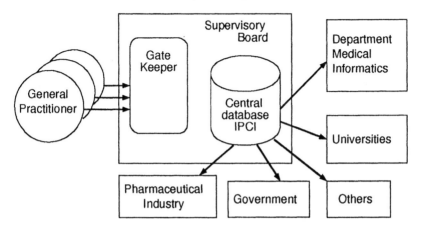

Figure 2. Organization of the drug-safety system

In the IPCI project we have to deal with many groups, all worried about at least their own interests. For example, the patients want to keep their privacy, the GPs want to continue delivering care, the researchers want to get more and reliable data, the government wants to control the costs of care and the pharmaceutical industry wants to survive by making more profit. Concerning the privacy issue, in the Netherlands, it is not allowed to collect data that may violate the privacy of a person. Anonymization is therefore an important condition. The doctors are responsible for the anonymization and transmission of the patient records. A gatekeeper is responsible for the anonymization of the doctors. All data collected in the central database of the IPCI project can not be used without explicit permission of a Supervisory Board. The majority of this board consists of doctors, so they can watch over their interests. To support the integrity of the research, a permission can only be given to start a specific research. Once a permission is given, the researcher may publish the results, no matter how disappointing they are to others.

Summarizing, when industries like IMS, or our own researchers from the department of Medical Informatics, or others like the government, want to investigate the (side-)effects of a drug, a permission of the supervisory board is needed and only anonymous data is available. All procedures and cooperative commitments are included in social and juridical contracts.

After the design, development and implementation of this drug-safety system, we want to evaluate the system. The importance of evaluation is at least threefold. Firstly, in the traditional method to develop a system, the system is usually divided into sub-systems with their own goals and purposes. After integration of those sub-systems the whole system must still be evaluated. Secondly, a system may have a great impact on persons involved in or affected by that system. Although it is difficult to point at people that may held responsible for such an assessment, the necessity to do this evaluation is less discussed. Thirdly, to communicate scientifically about systems and to compare them methodologically, the evaluation process itself must be part of a method which can be reproduced by other independent people. The importance of evaluation of hypotheses in natural science, for example, can be illustrated by a remark of Albert Einstein. When he recognized the freedom to generate a hypotheses, he raised the question: how to make a difference between a non-sense hypothesis and a hypothesis that somehow corresponds to phenomena in the natural world? His own answer was: only by testing the hypothesis (Einstein, 1934).

The movement of 'systems thinking' seems to be a suitable candidate to evaluate systems methodologically. First of all, because of its emphasis on scientific methods using analogies from natural sciences. Secondly, system thinkers are familiar with systems concepts and developing methods, because of its technical roots in operational research and cybernetics. Probably the most important reason to study systems thinking is the evolution of the systems paradigms. Not only technical systems, but social systems as well are the main points of view. Since we have an information system as a technical system and an organizational system as a social system, the methodology of systems thinking seems to be most suitable for the evaluation of our drug-safety system.

The question we want to address in this chapter is: does 'systems thinking' deliver adequate methods to evaluate systems of this kind?

2. METHODS

To answer this question, we want to investigate some methodologies of the systems thinking. The main criteria for this investigation of methodologies are: 1) What are the systems concepts and do they fit the drug-safety system? 2) What are the methods used to evaluate systems and can they be applied to the drug-safety system? 3) What are the boundaries of the evaluation phase of the methodology or what evaluation questions of the drug-safety system remain unaddressed? 4) Is it possible to generalize the evaluation of a specific methodology to an objective activity, executed by independent people? 5) What is the weakness of the methodology as such?

To address the different systems concepts, we will follow the evolution of systems thinking itself. "Until the 1970s," wrote Jackson (1991:4-5), "there was considerable agreement in the systems movement about how the notion of 'system' should be understood and applied. ... Summarizing greatly, it was assumed that systems of all types could be identified by empirical observation of reality and could be analyzed by essentially the same methods that had brought success in natural sciences." The methods used for producing and testing theory in the natural sciences are 'well understood' in the sense that the elements comprising the scientific method - observation, analysis, experiment, formulation of hypotheses, testing, development of theory - have been recognized for many centuries (Jackson, 1985). This type of systems thinking, using well understood methods, can be called the traditional system approach, later labeled as 'hard systems thinking'. Hard systems thinking can, following Checkland (1978), be seen as made up of three strands. These three strands are the methodologies of 'systems engineering', 'systems analysis' and 'operational research'. Each of these strands have several methods to evaluate systems. We will try to evaluate the drug-safety system with the engineering method.

During the 1970s and 1980s, the hard systems thinking became subject to increasing criticism. Limitations of the methodology appear. We will study whether these critics and limitations can be applied to the drug-safety system.

As a result of the obvious failings of hard systems thinking, especially when applied to social systems, alternative systems approaches were born, for example the 'soft systems thinking'. The debate between the hard and soft systems thinkers elicited fundamental differences in opinion. In fact, they rested upon different philosophical assumptions. In essence, they were based on different paradigms. For example, the (social) world is no longer seen as consisting of facts, events and causal relations, but the world is seen as being the creative construction of human beings. This means that systems are seen as the mental constructs of observers rather than as entities with an objective existence in the world. This also means that there are no problems 'out there' to be solved, but there may only be some people considering a situation as problematic. Traditional concepts like 'system', 'problem', 'conceptual model', 'solution' and also 'validation' and 'evaluation' were used in a total different way of thought. We will discuss the evaluation of the drug-safety system with respect to the method of the soft systems methodology.

The most radical critique on soft systems thinking came from social scientists. For example, Jackson used social theorists like Habermas to argue that if soft systems approaches are to be used wisely, there must also be an understanding that there are many social systems for which they are inappropriate. For such social systems there is a need for a more radical and critical approach to producing and verifying social systems theory and practice. Jackson proposed some methods in what he called: critical systems thinking. The absence of norms when coping with aspects of social reality such as conflict, contradiction, power, coercion and change gave rise to critical heuristics that may help to become self reflective with respect to normative implications. Since evaluation of systems is strongly associated with a variety of norms to be measured, we will also try to evaluate the drug-safety system in terms of the critical systems thinking.

In the present debate, many contributions consider philosophical or normative issues. A contribution which claims to do both is the 'multimodal systems thinking'. In this approach a philosophical structuralization of the world is given, in order to bring into account the variety of systems with their own singularity. For example, besides the technical and social systems, there are legal ·and ethical systems, mathematical and psychical systems and so on. Evaluation as systems assessment must address issues like the mutual impact between such systems. In the end we will also study how multimodal systems thinking may contribute to the evaluation of the drug-safety system.

3. RESULTS

3.1 Hard Systems Methodology: concepts and methods

To evaluate a system in the engineering tradition, means that we seek to assess the system's overall value. The important parts of the evaluation phase are verification and validation. Verification evaluates whether system specifications have been implemented correctly; it concerns 'building the system right'. Validation evaluates a system's performance and accuracy; it concerns 'building the right system'. An issue of verification is: are the usual methods and norms used in the right way? Does the system as a whole fulfill the technical and functional specifications in a reliable way? An issue of validation is: is the overall performance of the system after implementation in the concerning context fulfilling the needs of that context?

When applying to the drug-safety system we must first determine what kind of systems can be evaluated in this way. Since the focus is on technical systems, it is common use to distinguish several sub-systems, each of which can be verified separately. For example, the PMS-module, the transmission process, the inputsystem of the central database and the research programs are verified by independent testers. After several cycles they were proven to be good under testconditions. The functional and technical specifications are now fulfilled. The validation process, on the contrary, can not be applied to every subsystem, because it requires a user or a group of users. This means that sub-systems without users can not be validated, whereas the whole system with a mixed usergroup can be validated only by agreement of the group. Applied to the drug-safety system, the inputsystem of the central database can not be validated, because it is a fully automated process. The PMS-module is validated by a group of GPs and the database system including research tools by a scientific researcher (Visser, Vlug, Van der Lei and Stricker, 1996). The whole system including the protocols of anonymization, encryption and the role of the Supervisory Board is validated by the participating doctors. They reached an agreement on cooperation, resulted in a contract with the sponsoring industry.

3.2   Discussion on Hard Systems Approach: unanswered questions, objectivity and methodological weakness

Although the evaluation seems to be unproblematic, several questions are still unanswered. For example, the decision to validate the whole system only with participants is not evaluated. Man may argue that the 'affected' like patients, industrial competitors, pharmacists etc. also have the right to enter the validation process. Another issue is the limitation to evaluate only the existing system and not some invented systems. In other words, we evaluate the system as it is, not what it ought to be. A third problem is the absence of the evaluation of the assumptions made during design and implementation. For example, the safety of drugs is reduced to effectiveness of drugs, while the idea in patients mind that a drug is unsafe is a real danger for the pharmaceutical industry. With respect to this unpredictable aspect, it is not easy to assess the support the drug-safety system may give. It is obvious that the role of the media is important here, but they are not present in the drug-safety system at all. Questions remain unanswered, like: how fast can a permission be given by the board, when crisis management requires a report to disprove a statement in a newspaper?

The power of the hard systems evaluation is its objective reproducability. Independent programmers and computersystem-users are able to verify and validate the (sub)systems. In the verification phase, however, only the implementation of the specifications are verified. The completeness of the specifications, the weight of the particular specifications and the way in which the programmer reached the fulfillment of the specifications is beyond the technical scope. It is on this point some argue that the verification of professional work requires not only specification verification, but also verification of the way in which the profession was achieved. An objective tool to verify this professional attitude is a code of ethics for the programming professionals. Others argue this code of ethics belongs rather to systems ethics than to hard systems thinking.

The power and also the weakness of hard systems methodology is the focus on technical systems. The context of those systems is taken into account insofar it consists of users of those systems. The person-to-person communication and the context of a system-with-user are beyond the technical scope. The verification is also limited, because of the gap between user-requirements and technical specifications, the verification can only be done by programmers, not by users.

3.3   Soft Systems Methodology: concepts and methods

When hard systems methodologies are applied to social problems some failures may occur. The analysis of those failures have led to the conclusion that in social systems the subjective interpretations of a problematic situation are often more important than the discussion whether or not the problem is an objective problem in real world. The attempt to deal with those subjective problems methodologically, brings the soft systems thinking to the philosophical statement that there are only systems as interpretations: mental constructs of observers. Churchman, for example, stresses that the social world is perceived (and constructed) by men according to the particular world-views or Weltanschauungen which they hold (Churchman, 1968b). The concepts of 'system', 'problem' and 'solution', but also 'evaluation' will change radically. For Churchman 'objectivity' in social systems science is not to be sought by constructing theories and attempting to verify/falsify these by observation of the real world, because no observation of the world can be free of theoretical and metaphysical underpinning. According to Ackoff, there are no problems 'out there' to be solved. In any particular situation there can be many different definitions of what the problem is. The only objectivity is consensus arrived at through open debate. Objectivity is the social product of the open interaction of a wide variety of individual subjectivities (Ackoff, 1974). This participating criteria also applies to the concept of validation. The validation of a new or changed (social) system is based upon respect for the point of view and aims of all the persons affected by the intervention. Checkland, who proposed a methodology, uses no longer the evaluation as a separate phase. He suggests that a system is validated just by using the methodology. All questions one might still have concerning evaluation issues, point at the person who experiences a situation as problematic and for that case he developed a methodology (Checkland, 1985). Different persons or different opinions are brought together in a debate. The aim of the debate is to find some possible changes which meet two criteria: systematically desirable and culturally feasible in the particular situation in question. If this aim is reached, the solution is a validated one, because it is achieved methodologically.

When we try to apply this methodology to the drug-safety system, we focus on the problem of unsafety: who are the people feeling unhappy about drug-safety issues? Patients, doctors, responsible persons of industry and government? What about reporters? Suppose a reporter gets the feeling of a drug being unsafe and the reporter puts it in the newspaper. Many patients and doctors will get the feeling of unsafety too, justly or not. The organization of a debate with all the participating groups in order to manage this problem is not adequate. The particular pharmaceutical industry must react within a few days, otherwise they can go bankrupt. The drug-safety system is meant to support quickly further research on the conditions of the unsafety. The solution or the new situation for problems like this, will be an announcements concerning the specific drug: either it is withdrawn from the market, or it received a limited use, or the initial information about unsafety seemed to be groundless. However, this is not the problemsolving we like to evaluate. We like to evaluate a system searching for drug-(un)safety by using anonymized records of patients and resulting in reports that may have great consequences for industry or government. All those parties have their own interests, and must brought together in a debate. But they have at that moment no problem. We must convince them of a possible problem, before the debate starts. But the soft systems methodology is partly developed to overcome this expert-like influence.

3.4   Discussion on Soft Systems Approach: unanswered questions, objectivity and methodological weakness

The questions we want to answer are: who should we invite in the debate about the organization of the data-flow? Who should be responsible for the anonymized data? Who should be the owner of the anonymized data? What about privacy-issues people are not (yet) aware of: how can they play a role in the debate? What is right, in a situation of crisis management, when it is difficult to have an honest debate between the crisis perpetrator and the possible victims?

Exactly on this point Jackson formulates his critics on the soft systems approach. If validation is to depend upon the achievement of a true consensus among participating actors, care will have to be taken that the discussions that take place are free and unconstrained. The debate about change will have to conform to something like the model of 'communicative competence' proposed by Habermas (Jackson, 1985). This requires at least some norms about the debate, which cannot be discussed in the debate. Some kind of objectivity is needed to guarantee the openness of the subjectivity in the debate. The weakness of the soft systems methodology, concerning evaluation, is twofold: 1) strictly speaking, the action to change is not part of the methodology, because the focus is on 'subjective interpretations of situations'. This means, the evaluation of actions is not part of the soft systems methodology. More general, an objective reproducible validation is no longer part of the methodology. By the way, it is interesting to conclude that Jackson, who repeatedly stressed the importance of having an analogy with the successful method of the natural sciences, is not worried about the disappearance of validation as a separate phase. Probably this is the reason why he quotes Rescher so often. Rescher argues that a hypotheses does not become a scientific law because of repeated observation and experiment but only when it can be integrated into a systematic body of scientific knowledge - a cognitive system. The validation or justification of science proceeds along these lines. The scientific theses which emerge from the method are acceptable just because they are produced by an appropriate method - the systems/scientific method.

3.5 Critical Systems Methodology: concepts and methods

The critical approach, proposed by Jackson, is derived from the work of Habermas. This approach concerns validating work in social systems science. Habermas criticizes the system thinker Luhman on a major point: it is true that society can be seen as a system, but it is an environment to live in as well. Social reality may be taken as a social system, but it is a social practice too. Summarizing the points Jackson mentioned, Habermas can be used to put three critical remarks to the soft systems approach. Firstly, when the debate allows only interpretations of the participants, there are no norms to structure the debate on forehand. However, to guarantee the communicative competence an agreement on some conditions is necessary before even starting the debate. Secondly, Habermas (unlike soft system thinkers) regards the construction of explicit social theories as an essential part of any social systems science. Professional scientists must validate the knowledge theoretically. Thirdly, this theoretical validation is not sufficient. The knowledge must also be validated by social actors. Only if the theory helps these actors to attain self-understanding and they recognize in it an acceptable account of their situation, the theory can be said to be validated. Whereas the soft systems approach presupposes that debaters are aware of their deepest problems, Habermas provides a psycho-analytical method to attain self-understanding.

Despite all the valuable notions mentioned, no cohesive method of thinking appears. Focussing on the conditions for a perfect communication, there is still the problem how to communicate in the case of certain opponents who are, for example, in the grip of ideology. Jackson replies that in such cases it will be necessary temporally to break off the dialogue and to engage in political struggle against these opponents (Jackson, 1985). However, the attractive ideas about evaluation are than no longer justified. Critical systems thinking seems to stick to an antithetical relation with the soft systems thinking.

An attempt to make some thetical steps, can be considered in Ulrichs contributions. He provides a systematic way to explore Habermas' distinction between theory and practice. By adding the practical philosophy to the systems philosophy, Ulrich is able to bridge the gap between systems thinking and the life-practical concerns of politics, morality, religion and aesthetics. Influenced by Kant, Ulrich proposes to switch from systems science to systems rationality and to distinguish between a theoretical rationality and a practical rationality. The validation of hypotheses depends on (critical) rationality, but the task of the practical rationality differs from the task of the theoretical rationality. While theoretical reason must decide on disputed claims regarding the empirical validity of theoretical propositions (hypotheses), practical reason must decide on disputed claims concerning the normative validity of practical propositions (assertions of norms, recommendations for action). In order to reflect and debate systematically on the normative implications of systems designs, Ulrich needs both the idea of practical reason as a critical standard against which to examine the instrumental rationality, and the systems idea as a critical reminder to reflect on those implications of design which reach beyond the limited context of application - their whole-systems implications (Ulrich, 1988). To summarize some important notions, Ulrich comes up with: firstly all methodologies have their own limitations. For example, no standpoint, not even the most comprehensive systems approach, is ever sufficient in itself to validate its own implications. Instead of Jackson's search for the conditions of unconstrained discussion it is better to deal critically with conditions of imperfect rationality. Secondly, critically normative reflection must not remain extrinsic to systems thinking and systems practice, rather it must become an intrinsic part of our understanding of systems rationality. Critical heuristic may help to become self-respective with respect to the normative implications of any standard of rationality. For example, to lay open the limitations and normative implications of instrumental rationality, to recognize the one-dimensionality of strategic action and to systematically trace the normative implications of systems designs.

3.6   Discussion on Critical Systems Approach: unanswered questions, objectivity and methodological weakness

Concerning the evaluation of the drug-safety system, the critical systems rationality provides a framework in which all our questions have a legitimate place. It even improves the evaluation of the other approaches. In the hard systems approach, for example, when the system is verified and validated, thus when we know what is done is good, it is now possible to ask: what should be done? And how should it be done? Also the decision to determine the systems boundary is an action to be evaluated. In the soft systems approach we now must recognize that there is no explicit directive that prevent expert domination by discussing the mutual cooperation. Furthermore, the decision who may participate in the discussion appears to be an ethical one. Concerning the question how to manage a crisis, it is now obvious that this question not only belongs to systems thinking, but to systems practice as well.

The power of the critical systems thinking as a methodology is the acknowledgement of the normative side of a debate and the systems practice in relation to systems thinking. However, whereas the hard systems approach have its own methodology, and the humanities have their hermeneutic method, the systems practice have not established an objectively (or intersubjectively) reproducible way of ensuring rational practical discourse on disputed norms of action. On the other hand, the heuristical methods available in critical systems methodology can be used as a framework to guide the design and evaluation. However, though we may have a conceptual framework in which all evaluation issues can be addressed, there is no method ensuring the completeness of these issues. For example, Churchman enumerates the 'enemies of systems rationality': politics, morality, religion and aesthetics, but are these the only ones?

3.7   Multimodal Systems Thinking: concepts and methods

A detailed framework of systems theory and systems practice, critically addressing normative issues is the multimodal systems approach. The framework is build by De Raadt (1994) along two dimensions: a string of interacting systems in which we live lies on the systemic axis. For example, orchards, families, trades, dance, companies, schools, etc. On the modal axis a number of modalities of the world is layered: the credal, ethical, juridical, aesthetic, economic, operational, social, epistemic, informatory, historical, logical, psychic, biotic, physical, kinetic, spatial and numeric (Dooyeweerd, 1958). These modalities are put forward by the philosopher Dooyeweerd, by observing the reality and determining which modalities are irreducible to each other. The multiplicity of these modalities is evident in the variety of specialized sciences. For example, the credal, ethical and social modalities are respectively studied by theology, ethics, and sociology. More important than the philosophy of the modalities, is the practice of acting wisely on the systemic axis. Each act reflects all the modalities, but the amount of freedom to act is dependent of the kind of modality. The numeric modality is mostly commanded by nature, whereas the credal aspect is less commanded by nature and mostly by the freedom of the acting person. Though the evaluation process is not explicitly described, we will propose to formulate some evaluation issues that may be supported by the multimodal framework. Firstly, the multimodal approach seems to deliver a complete enumeration of modalities. Evaluation may start by determining what the main modality of the system is and for that modality the amount of natural norms. Secondly, justice is done to the intuition that physical norms differs from juridical norms. Physical norms seems to be more objective (everybody experienced the law of gravity, without trying to overcome it), whereas juridical norms seems to be more subjective and disputable. Thirdly, systems may be divided into sub-systems, each of which may have an own purpose on one of the modalities.

Applied to the evaluation of the drug-safety system, the multimodal approach can be used to improve the design of the hard systems methodology. In trying to find an exhaustive list of specifications, the list of modalities may help. In the evaluation process it is now possible to explicitly investigate whether some modalities are (or should be) more important than may be expected based on the purpose of the system. For example, the PMS-module has its main purpose on the epistemic modality, because the recording of the events must be coded and completed. The main purpose of the GP practice is the ethical: the care for the patient. When the PMS-module becomes dominant in the consultation, the main purpose (delivering care) may be overwhelmed by the epistemic modality. An evaluation according to the multimodal systems approach supports a systematic way of eliciting such ethical issues. The multimodal approach can also be used to improve the soft systems methodology by searching for norms preliminary to the debate and providing a framework of modalities to guide the awareness during the debate. The critical methodology can be improved by searching not only for norms concerning the way of debating, but also for norms concerning the systems themselves. This is important in the evaluation phase of the systems methodology. These norms are not disputable on the basis of (power-dependent) opinions in a debate, but they are disputable on the basis of the question: what is the multimodal purpose of the system and what is its qualifying modality? The answers to these questions direct the design, for example the choice which methods to use, and the evaluation, for example the importance of each of the norms.

3.8   Discussion on Multimodal Systems Approach: unanswered questions, objectivity and methodological weakness

The structuralization of the world, based on a 20th-century philosophy is disputable. Even when we assume that this philosophy is valid for 20th-century problems, than we can still ask: Why are the epistemic and informatory modalities irreducible and is music, for example, not a separate modality? The list of modalities may reflect the observations of reality, but there is no evaluation of what reality ought to be. The advantage for the design and evaluation process is to have a 'checklist' in order to manage the creative phases of the methodology. The course of the amount of natural norms through the modalities is convincing, but how to gain knowledge about the content of the norms? What about the integration of conflicting norms? It seems to me that some powerful concepts are given, but they are not yet usable for the methodological evaluation of systems.

4.   CONCLUSIONS AND RECOMMENDATIONS

Evaluation of a system needs criteria whether a system is good or not. Depending on the problem or problematic situation the system have to cope with, these criteria are more or less disputable. The more hard problems are, the less disputable the criteria seem to be. When we search for an evaluation which is reproducible by others, we prefer less disputable criteria, because the more objective criteria are, the more easy evaluation can be done by independent persons. When the criteria are disputable, the evaluation changes. It is possible to evaluate the system, given the consensus among criteria generated by the system developers, and it is possible to evaluate the consensus itself.

Hard systems methodology is most suitable for well-structured problems, the criteria are well-described and the evaluation of those technical systems can be reproduced by others. However, most systems are not entirely technical. Systems may have technical sub-systems, or systems may have more relevant perspectives than the technical one, and systems may be not technical at all.

Soft systems methodology is most suitable for situations, experienced by different people as problematic. Since the methodology is limited to interpretations of situations, there is no situation as such to be evaluated. If the way in which a consensus is reached, becomes part of the evaluation, then the method instead of the system is evaluated. Criteria to evaluate the method are not generated in a debate.

Critical systems methodology is most suitable for systems with a hard and a soft side. The methods of both hard and soft systems methodologies can be used, as long as the boundaries of the particular method is recognized. The limits of subjectivism can be overcome by a normative approach of the debate. However, a consistent way to evaluate the balance between interpretation and action, between norms and the obedience to them, is not available.

Multimodal systems methodology is most suitable for systems that are complex on two levels: they can be analyzed in many subsystems and they can be considered from different relevant perspectives (or modalities). Since there is a structural relation between the world as such and our multimodal interpretation of it, the evaluation of a system can be reproduced by others based on set of norms discovered by others, while the evaluation of the interpretations must be as multimodal as the situation itself. However, methods to search for the right norms are yet to be developed.

Chapter 10

MULTIMODAL INVESTIGATION OF TECHNOLOGY-AIDED HUMAN PRACTICE IN BUSINESS OPERATIONS

Darek M. Eriksson

"What's emerging from the pattern of my own life is the belief that the crisis is being caused by the inadequacy of existing forms of thought to cope with the situation. It can't be solved by rational means because the rationality itself is the source of the problem."

Robert M. Pirsig (1981:149)

1. INTRODUCTION

This article presents a study where an implementation of a new business process model supported with a new computerised information system has taken place causing some unpredicted and unwanted consequences. This case is investigated with Multimodal Theory, part of a recent school of thought called Multimodal Systems Thinking. Reflections upon the employment of these theories provide some heuristics for further development and use of this particular kind of systems thinking. The following gives an introduction to the topic. The communities of information systems, systems thinking and management science, along with several others, have operated in a kind of internal intellectual conflict, having two seemingly opposing meta-theoretical positions. One is the positivistic foundation, as inherited from the natural sciences, while the second is the hermeneutic foundation, as inherited from the social sciences - for this discussion see for example Checkland (1981) or Iivari, Hirschheim and Klein (1998). In the recent search for new and alternative theoretical and meta-theoretical foundations, in order to solve this tension, the so-called Frankfurt School of social studies and particularly its younger representative Habermas' (1984, 1987) works on Critical theory, have received significant momentum. Examples of its influence in the information systems community include Lyytinen (1986), Ngwenyama (1987), Hirschheim and Klein (1989), Lyytinen (1992), Klein and Hirschheim (1993), Kendall and Avison (1993), Hirschheim and Klein (1994), and in the systems thinking community are Ulrich (1983, 1987), Flood and Jackson (1991a, 1991b), Flood and Romm (1996a, 1996b).

Among other things, Critical theory offers a broader notion of rationality, not limiting itself to the goal-oriented or technical rationality, as the positivistic foundation tends to do, nor to the communicative rationality as the hermeneutic foundation tends to. Because of its Kantian foundation, Habermasian thought accounts for three kinds of rationality or inquiry: empirical analytical sciences, which focus on instrumental-reason and provide nomological causal knowledge that aims at prediction and control of nature, historic-hermeneutic sciences, which provide the practical understanding of other human beings, and critically-oriented sciences, such as psychoanalysis and critical social theory, which provide emancipatory interest in freedom and overcoming unconscious compulsion. The critical foundation offers a complementary notion of rationality that tries to bridge the gulf between the positivistic and the hermeneutic positions.

Seemingly independently of each other, Ivanov (1996), in his critique of Hirschheim, Klein and Lyytinen's (1996) attempt to map the information systems development approaches, Strijbos' (1995), in his evaluation of systems thinking schools, and Eriksson (1998), in his investigation of systems thinking positions, have all criticised the Habermasian foundation. As Strijbos (1995:374) put it: "Science as an instrument of control is subjected to criticism by 'Critical Systems Thinking' but this critical thinking remains subject itself to an autonomous rationality." The three critics all, in their own way, propose an alternative foundation to the Habermasian, in the so-called Multimodal Systems Thinking that is founded on Dooyeweerd's (1894-1977) Cosmonomic philosophy and social theory. Even though Multimodal Systems Thinking has firmly established theoretical and meta-theoretical foundations it lacks a wide experience of application, limited to a handful of empirical studies (see for example Bergvall-Kåreborn and Grahn, 1996a, 1996b; Mirijamdotter, 1998; Bergvall-Kåreborn, 2000). To help to remedy this limitation, this article presents a study, employing the Multimodal Theory that is part of Multimodal Systems Thinking, of an implementation of a new business process model supported by a new computerised information system, which had some unpredicted and unwanted consequences.

The next section outlines Multimodal Systems Thinking in general and Multimodal Theory in particular. The following section describes the investigated system with its two situations: before and after its re-design. After this, we undertake an analysis of the system, using Multimodal Theory. Finally, we reflect on and discuss the analysis, its advantages and limitations, draw conclusions and suggest issues for further research.

2. MULTIMODAL SYSTEMS THINKING

This section starts with a presentation of Multimodal Systems Thinking in relation to the three more established alternative scientific paradigms in systems design. Then a very brief mentioning of four dialects within Multimodal Systems Thinking is given that allocates this particular study's position. Next follows a short overview of the Dooyeweerdian theories. This is then followed by the main part of this section that presents the Multimodal Theory, which is employed in the present study.

2.1 Multimodal Systems Thinking in the context of three established system thinking paradigms

Multimodal Systems Thinking (MST) emerged at the end of 1980's and the beginning of 1990's as an alternative scientific paradigm for systems design and management (see for instance De Raadt, 1989, 1991), both in general and in relation to information systems (De Raadt, 1991; Winfield, Basden and Cresswell, 1995; see also chapters 2,4). MST arose very much from the dissatisfaction with, and as an alternative to, the two well established scientific paradigms, the positivistic and the hermeneutic, and the third currently establishing itself, the critical.

The main accusation towards the positivistic approaches delivered by the hermeneutic position was concerning the nature of reality and knowledge, the nature of human beings and also the scientific methodology of inquiry. The positivist's position assumes that there is a reality independent of human thought and that an objective knowledge of that reality may be acquired. It also assumes that humanity and its actions are deterministic and a product of the external world. Positivist methodology is nomothetic, aiming for description, explanation and prediction of the world in terms of deterministic and stochastic behaviour. The hermeneutic position considers the positivist notions a misconception that leads to a reduced knowledge of the studied phenomena. Its position is that reality is a product of individual consciousness, and that knowledge is subjective, based on experience, insight and essentially of personal nature. The hermeneutic position also assumes that humans are voluntaristic or teleological, have creative role and free will and create their environment. Its methodology is ideographic, aiming to understand the way an individual creates, modifies and interprets the world. (For more on this see, for example, Checkland, 1981, Flood and Jackson, 1991a, Flood and Carson, 1993). Though sharing the hermeneutic critique of the positivistic approaches, critical approaches accused the hermeneutic position, among other things, of isolating itself from the positivistic position instead of being complementary, of not being self-critical and as a consequence for its relativistic tendency, and of being normatively - and thus ethically - ignorant and therefore regulative rather than emancipatory (see for example Flood and Jackson, 1991a, Flood and Romm, 1996). Multimodal Systems Thinking, on the other hand, criticises the critical position for accepting and promoting the intellectual supremacy of rationality. Strijbos (1995:374) articulates the argument as follows:

> "Basic to this critique is the insight that rationality may not be separated from reality as it is given and from the insight that rationality as human reason is embedded within a supra-subjective and supra-arbitrary normative order of reality. This normative order preceded reason and every discussion of rationality, and it is therefore of fundamental importance for determining the status of reason and of science."

Eriksson (1998) presents a critique with similar consequences, elaborated in chapter 13. Elaborating the Dooyeweerdian theory that ground-motives necessarily founds all theories, he concludes that the Kantian and thus also approaches founded on Habermas have an inherent and unresolved theoretical conflict that makes its normative guidance blind and reduced to intellectual reasoning and argumentation.

2.2   Various dialects of Multimodal Systems Thinking: position allocation

The development of Multimodal Systems Thinking has mainly taken place in the context of the *Centre for Technology and Social Systems*[1]. It is possible to distinguish several sub-developments within this school of thought. One is a combination of Dooyeweerdian theory with cybernetics (Ashby, 1960; Beer, 1979, 1981) and General Systems Theory (Von

---

[1] See Preface for an explanation of this Centre, now called the Centre for Philosophy, Technology and Social Systems.

Bertalanffy, 1968; Boulding, 1956), among others, as pursued by Donald and Veronica de Raadt (see for example De Raadt, 1991, 1998, 2000). A second is the combination of Dooyeweerdian theory with Soft Systems Methodology (Checkland and Scholes, 1990a) as exercised by Bergvall-Kåreborn and Grahn (1996a, 1996b), Mirijamdotter (1998) and Bergvall-Kåreborn (2000); see chapters 3,5. Another attempt is the combination of Dooyeweerdian theories with various information technology domains, particularly Expert Systems and Multimedia; see Winfield, et. al. (1995) and chapters 2, 4, 11. Still another attempt is the combination of Dooyewerdian approach with Operational Research in general as performed by Strijbos (1995); this approach was recently re-labelled to 'Disclosive Systems Thinking' (Strijbos, 2000); see chapter 14. Even though all these attempts share the same Dooyeweerdian foundation and are very much interconnected there are differences in their approaches and interpretations, and also in the type of combinations they attempt with other, non-Dooyeweerdian theories. Such diversity can only be healthy when a new scientific school is establishing itself, as was the case with the early Frankfurt School and its Critical Theory (Burill, 1987:8). This present study attempts to use Dooyeweerdian theories with basic business process modelling techniques.

2.3   A brief map of Dooyeweerdian theories

Herman Dooyeweerd (1894-1977) was a Dutch lawyer and philosopher. His philosophy was, among other things, a reaction against the then dominant neo-Kantian trend in continental thinking; hence the title of his monumental work *A New Critique of Theoretical Thought* (Dooyeweerd (1955-58). The result of his work may be organised into five distinct, yet interrelated, domains of thought: the theory of religious ground-motives, the modal theory, the theory of time, the entity theory or theory of individual structures, and the social theory. The present investigation uses the modal theory as a tool of investigation.[2]

2.4   Multimodal Theory

Multimodal Theory emerged out of Dooyeweerd's comprehensive studies of theoretical thought and its relation to human reality.

---

[2] We have elsewhere employed the theory of religious ground-motives as a tool for investigation; see chapter 13 and Eriksson (1998).

Dooyeweerd maintained that our thought is based upon and bound to our experience and that this experience exhibited a number of distinct modalities (or levels, or aspects, or dimensions, or spheres) of organisation or law. Hence, a modality emerges out of human interaction with her reality (which include both perceptions and conceptions), and it is a particular type of knowledge that has its own unique and distinct characteristics. Dooyeweerd proposed fifteen modalities, in the following order: arithmetic, spatial, kinematic, physical, biotic, sensitive or psychic, logical, historical, lingual, social, economic, aesthetic, juridical, ethical and pistic, see Table 1 for an overview.

Table 1. The modalities and their related nuclei.

MODALITY	NUCLEUS
Credal	Faith
Ethical	Love
Juridical	Justice
Aesthetic	Harmony
Economic	Frugality
Social	Social intercourse
Informatory	Symbolic representation
Historic	Formative power
Analytic	Distinction
Psychic	Feeling
Biotic	Vitality
Physical	Energy
Kinetic	Motion
Spatial	Continuous extension
Numerical	Discrete quantity

It is important to note, however, that Dooyeweerd's intention was never to construct a fully comprehensive and exclusive map of human experiences; it is a proposal and he welcomed suggestions for modification, which has been exercised for example by De Raadt (1998).

The element that distinguishes one modality from another is its respective nucleus or kernel, which makes each modality meaningful, unique and irreducible to others; it provides a hub toward which entire modal order is aimed. Further, the division of the modal continuum into discrete modalities is made for diagnostic reasons; in practice they never appear isolated but rather in an inseparable intertwinement. Each modality is founded upon the preceding ones, so the given order of modalities is not accidental. This allows a certain degree of homomorphism between modalities, the degree of which depends on the distance between modalities. This makes it possible to transduce knowledge between modalities. For example, social scientists often express aspects of social behaviour (Social modality) in terms of quantitative measures (Numeric modality). Mathematical manipulation of the representations and the conclusions derived rest upon the laws of the Numeric modality and not on those of the Social modality. Therefore while mathematically valid they may not be valid in the social sphere (De Raadt, 1998).

3.   THE INVESTIGATED SYSTEM

3.1   Organisational background

The business organisation of this investigation was a vegetable, fruit and flowers sales department at a wholesale dealer that belonged to one of the largest food distribution company in Sweden, and was located in a medium size city in the south of the northern Sweden. The company had for some years invested significant financial resourses in the development of a new nationwide information system (IS). The motive for that development was both to create new kinds of business processes, which it was not really possible to actualise without information technology support, and to co-ordinate the existing business processes and make them more efficient, mainly by automating the existing manual routines. The present case study investigated one particular re-design of a sales process.

3.2   The sales process prior to its re-design

Fig. 1 shows a business model of the sales process prior to its re-design. The wholesale dealer stored goods that were ordered by local food-selling stores in the area, which sold the goods to the local public. The vegetable, fruit and flowers sales department had six salesmen, who handled selling of the goods to the local stores.

This shows a typical ordering scenario: a salesman from the wholesale dealer rings a buyer at a local food-selling store. Sometimes a buyer rings a salesman. Then a dialogue takes place over what the content of the order should be. The content of this dialogue exhibits a certain complexity and will be discussed further below. This dialogue leads the buyer to take a decision about the content of his or her order. The salesman that receives this order enters it into the order-IS, which sends to the stock department a request for a delivery of this order. There the order is prepared for its delivery and then delivered to the local food-selling store that ordered it.

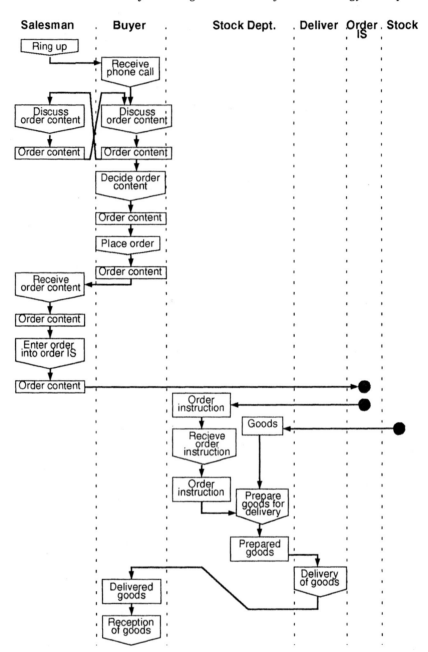

Figure 1. The sales process prior to its re-design.

3.3 Re-designed sales process

Fig. 2 shows a business model of the sales process after its re-design. In this process the buyer makes a decision about the content of an order and then places the order through an IS terminal. This order is transferred directly through the new order-IS to the stock department of the wholesale dealer. When the stock department receives the order they prepare it and deliver the goods to the local food-selling store that placed that order.

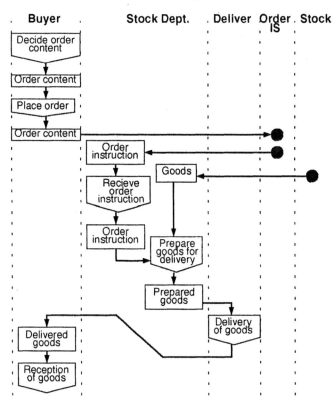

Figure 2. The sales process after its re-design.

3.4 Motivation for the sales process re-design

The motivation for the sale process re-design was mainly to increase efficiency. More specifically, a reduction of operational costs of the wholesale dealer by elimination of the costs of six salesmen positions, and also an anticipated increase of ordering access, speed and accuracy that potentially would lead to an increase of ordering volume. All this aimed to increase the profitability of the wholesale dealer. The cost of the process re-design and of the new IS, together with their implementation, were not considered directly here but financed by a central national budget of the whole food distribution company.

3.5   Juxtaposition of the two versions of the sales process

The difference between the two versions of the sales process, summarised in Table 2, can be described in typical business process re-engineering terms. In the sales process prior to its re-design the number of actors involved was six, a buyer, a salesman, the stock department, the goods deliverer, an order-IS, and a stock. The number of activities required to complete this process was twelve while the number of hand-overs between different actors was eight. In the re-designed sales process the number of involved actors was reduced to five by removal of the salesman. The dialogue between buyer and salesman, that preceded a buyer's decision about an order's content, was eliminated. The number of hand-overs between different actors was reduced from eight to five. The time required to place an order decreased significantly. Finally, the running costs of the new sales process implied a significant decrease, mainly by elimination of the costs of hiring six salesmen and those associated with their operations.

Table 2. Juxtaposition between the sales process before and after its re-design.

Comparison criteria	Sales process status	
	Prior to its re-design	After its re-design
Number of involved actors required to complete the process:	6	5
Number of involved activities required to complete the process:	12	6
Number of hand-overs between actors:	8	5
Salesman-Buyer dialogue:	Included	Excluded
Required time to complete the process:	Standard	Reduced significantly
Costs of six salesmen to complete the process:	Included	Excluded

3.6   Some unanticipated consequences of the sales process re-design

Our investigation observed the immediate consequences that the above described process re-design implied. As anticipated, the salesmen abolition reduced the operative running costs of the wholesale dealer. However, the process re-design led also to unanticipated and unwanted consequences.

The ordering behaviour of buyers of the local food-selling stores changed. The total amount of goods that they ordered per unit time decreased significantly, leading to a reduction in the total turn over volume, which in turn led to a significant profit reduction for the wholesale dealer. In abstract monetary units, the profit reduction was 45 money units per time unit while the reduced costs for the abolished salesmen was 30 money units per time unit. This suggests that the main reason for the re-design - to increase the overall profitability by exchanging the buyer-salesman-IS network into a buyer-IS network - failed.[3] In the following section an analysis of the situation is performed with the help of Multimodal Theory.

4. ANALYSIS OF THE INVESTIGATED SYSTEM

The central question of this analysis is: Why did the sales process re-design lead to a changed ordering behaviour of the buyers? In order to tackle this issue we need to set a focus for the analysis. One of the main differences between the two versions of the sales process is that in the process prior to its re-design there was a relation between two actors, which was eliminated in the re-design, the relation between buyer and salesman. Another difference is that in the process prior to its re-design there was a relation between salesman and order-IS while in the re-designed process this relation was exchanged for a relation between buyer and order-IS. This analysis assumes that the relation between the buyer actor and the salesman actor provides insight into our question above.

4.1 Further description of the buyer-salesman relation

A buyer ordered normally from the same salesman over a period of some years. This resulted in a certain relationship between these two that also led to an emergence of certain norms in this relation, a sub-culture. A typical content and structure of the ordering dialogue that a buyer and a salesman carried out through the telephone is illustrated in Fig. 3.

---

[3] The assumption that it was the sales process re-design that caused the decrease in profit may always be challenged. For example, there could have been other parallel ongoing activities that impacted the sales behaviour, such as new competition on the market. These have not been taken into consideration in this investigation. Further, this particular study was limited in its time scope of investigation. It is thus possible to speculate that after some time, the sales behaviour would reach the previous sales level. This point was also beyond the scope of this investigation.

	ACTOR	STATEMENT
Part I	**Buyer:**	Hello John! How are you?
	Salesman:	Hello Peter! Thank you, I am just fine.
	Buyer:	How are your wife and the kids?
	Salesman:	Not bad at all! Although Tom has been sick for a week. How is Diana?
	Buyer:	She is just fine but rather busy with her new job.

Part II	**Salesman:**	What would you like to have today?
	Buyer:	I'd like 20 boxes of oranges and 40 boxes of bananas.
	Salesman:	Peter! I have been down to the store. The oranges are fine but the bananas are so-so. I would recomment that you order only 10 boxes of bananas. The apples are very nice, however, and I can give you a special price. Why not order 30 boxes of apples?

Figure 3. A typical content and structure of the ordering dialogue that a buyer and a salesman carried out through a telephone. Part I of the illustrated dialogue represents actualisation of a personal relation while Part II actualises a professional relation.

Further, most buyers used to visit their salesman at the sales department with a varying frequency, some of them as often as once a week. One of the main aims of that visit for a buyer was to inquire about the present situation and the plans for the near future about types of goods that were available and that would be delivered to the wholesale dealer. Also the buyers used to inspect the goods stored in the stock department in order to assess the quality for themselves. All this information was very useful for the buyers' planning. These visits were also used to deliver complaints when such a need arose.

4.2   Analysis of the buyer-salesman relation

4.2.1   Identification of modal norms in the sales process prior to its re-design

This modal analysis starts with a description of the modal norms of the investigated actor relation prior to the process re-design.

The Social modality: A typical buyer-salesman ordering dialogue that was carried out through a telephone had two main parts (see Fig. 3). The first part of the conversation actualised a personal relation between the two actors where the norm required asking each other about their private situation and families. Our investigation showed that this information was important for a salesman since it gave him a description of "the mental state of a buyer and his mode of being" (quotation from an interview), which helped him to set the mode, or strategy, for the sales dialogue.

The second part of the dialogue actualised a professional relation between the buyer and the salesman. The norm here was that the buyer started this dialogue by informing the salesman about his needs. The salesman then reflected upon its content and gave a response. This response could typically be either a full agreement with the buyer, a proposal to change the quantity of the mentioned goods to be ordered and/or a suggestion to order other goods not mentioned in the first order statement. This was typically followed by a discussion where the salesman gave the reason behind his offer. Eventually, the discussion led to a consensus, with agreement on the content of the order.

Historic modality: The relation between a particular buyer and his/her salesman was to a certain degree formed or determined by their previous mutual experience of meeting each other and their conversations, which could mean sharing their private and professional problems and successes.

Spatial modality: Typically there was a varying spatial distance between a salesman in an office at the wholesale dealer and a buyer at a local food-selling store. This distance affected the kind of social intercourse between them. Typically a longer spatial distance meant that the buyer visited the salesman less often, and could not so easily see the goods him/herself but had to be informed by the salesman. Moreover, the distance made them interact mostly through a telephone, which also set conditions on the kind of interaction that was possible.

Informatory modality: The salesman informed the buyer about goods that were available, and their qualities and quantities. This information was typically conveyed through the telephone, or through face-to-face discussion during visits.

Kinematic modality: The buyers had to move from their respective local food-selling stores to the wholesale dealer in order to visit their salesman to inspect the goods in the storage, among other things. This norm of movement and its frequency was very much determined by the spatial distance between a buyer and his/her salesman.

Psychic modality: Three norms have been identified in the Psychic modality. First, people typically need social intercourse with other people in order to reach a state of mental wellbeing. This is especially the case for the so-called extrovert personalities that salesmen and buyers typically are. Second, salesmen were motivated by their salaries being based partly on the sales they achieved. Finally, a buyer's motivation was grounded in the dialogue with and offer provided by the salesman.

Economic modality: Economic modality may be considered here as the qualifying one (see chapter 3), i.e. the one that motivates the very existence of the inter-actor relation. This relation was about trading, and would not be actualised without it. Each salesman had a goal of a defined selling volume, upon which their salaries were partly based. If a salesman managed to sell more than the set goal, s/he received a bonus.

Ethical and Credal modalities: A salesman would never provide misleading or untrue information about, for example, the quality of the ordered goods. This ethical relation was in turn based upon a professional trust in each other that was a norm that belongs to the Credal modality.

4.2.2   Identification of inter-modal norm relations in the sales process prior to its re-design

The following inter-modal relations were identified; see Table 3.

The personal and professional norms of Social modality gave rise to, and were supported by, the professional norm of Ethical modality. More specifically, a salesman's advice or proposition to a buyer, about the content of an order, was considered to be correct and not misleading.

The norm not to misguide a buyer, in the Ethical modality, was supported by the established trust being a norm of the Credal modality.

The professional and personal norms actualized in the social relation between a buyer and a salesman (Social modality) were partly formed by the mutual past experience of their relation (Historic modality).

The psychic norm of mental wellbeing (Psychic modality) was supported by the social intercourse norms of the Social modality.

The economic norm of maximising selling of the goods (Economic modality), was supported by the motivational norm of the Psychic modality, which in turn was triggered by the salary norm present in the Economic modality.

The spatial distance (Spatial modality) conditioned the norms for social intercourse present (Social modality). The same spatial distance norm also determined the physical movement of the buyers, who moved in order to visit their buyer and the wholesale dealer (a Kinematic norm).

Informatory norms were conditioned partly by the Spatial norms, by whether exchanges were conducted through telephone or face-to-face.

Table 3. Result of modal norm analysis and the identification of inter-modal relations of the sales process prior to its re-design.

Modality	Norms	Relations
Credal:	A buyer's trust in a salesman to convey true information.	
Ethical:	A salesman would not provide untrue information, for example, about the quality of the ordered goods.	
Juridical:	A legal order contract was established in the buyer-salesman order dialogue.	
Aesthetic:	No relevant norms were found.	
Economic:	A salesman's salary was based upon the volume of money she or he could sell for. When selling more than the defined goals the salesman received a special bonus.	
Social:	Private role: to ask about each other's private situation and their families. Professional role: the buyer informed the salesman about his needs. The salesman agreed or proposed modification of quantity of the required goods and/or suggested ordering other goods.	
Informatory:	The information was conveyed mainly through the telephone, and occasionally through face-to-face dialogue.	
Historic:	A buyer-salesman relation was formed by several years of interaction, which could include sharing their private/professional problems and successes.	
Analytic:	No relevant norms were found.	
Psychic:	* People need social intercourse with other people in order to reach a mental state of wellbeing, especially relevant for extrovert personalities, which buyers and salesmen tend to be. * Each salesman was motivated to sell as much as she or he could yet following the established norms between the salesman and a buyer.	
Biotic:	No relevant norms were found.	
Physical:	No relevant norms were found.	
Kinematic:	The buyers did move physically in order to visit their buyer and the wholesale dealer.	
Spatial:	Spatial distance between buyer and salesman led to: * shorter distance implied more frequent buyer visits of salesman, longer distance implied less frequent buyer visits of salesman; * a buyer could not see himself the quality of goods that he or she ordered but had to be informed by its salesman; * that most of the buyer-salesman social intercourse to be performed through a telephone which in turn set conditions on the type and content of dialogue they performed.	
Numerical:	No relevant norms were found	

4.2.3 Identification of consequences of sales process re-design on the established modal norms

In this section the main questions are: (a) Which established norms and inter-modal norm relations in the sales process prior to its re-design were impacted by the process re-design? (b) What were the consequences of changes in the modal norms and inter-modal norm relations?

All the previously identified modal norms that constituted the buyer-salesman relation, except for the Kinematic and the Spatial, became modified or exterminated when the salesman was excluded from the new sales process. Thus the trust norm in the Credal modality as was established between a buyer and a salesman became exterminated. The same happened to the Ethical, Economic, Social, Historic, and Psychic modal norms. The Kinematic norm of buyer's movement to the wholesale dealer remained intact as was the Spatial norm of distance between a buyer and the wholesale dealer, where the salesman used to be.

Starting with the Credal norm, the trust that governed the buyer-salesman relation disappeared. The buyer had to trust the new IS but the relation between two humans is fundamentally different than between a human and a machine. Among other things, machines can not be assigned the property of responsibility. Trust in an IS is of a different type and takes time to build up. We find it likely that the extermination of the Credal trust norm impacted buyers' buying behaviour so that they became more careful and were motivated to order less than previously.

Moving to the Ethical norm, a very similar conclusion is reasonable. The extermination of the ethical norm that only correct and not misleading information should be provided to the buyer, made the buyers act more carefully and hence order less amount of goods.

The Economic modality's norm of the buyer-salesman relation was that a salesman's salary was based upon the amount he or she was able to sell. The extermination of this norm implied that the motivation (within the Psychic modality) for selling more disappeared. We conclude that this affected the buyers' buying behaviour so that they ordered less.

Considering the Social modality, both the identified private role and professional role norms became exterminated. This must be seen as the crucial extermination, because without buyer-salesman social intercourse most of the other modal norms could no longer be maintained. Since the social norms governed and conditioned the selling process, a reasonable conclusion is that their disappearance impacted the buyers' buying behaviour negatively in respect to the amount of ordered goods.

The norms of the Historic modality conditioned the buyer-salesman relation. The removal of these implied that they could no longer condition the buying behaviour. Indeed the buyers had no history with the new IS.

The extermination of the established Psychic norms implied that there was no motivational property, between a buyer and the IS, that drove and promoted the selling process. Considering the psychic norm of social interaction, it is reasonable to consider that its satisfaction was fulfilled by interaction with other people.

Table 4 summarizes the modal norm extermination as discussed above.

Table 4. Result of modal norm analysis of the sales process after its re-design.

Modality	Norms
Credal:	Exterminated
Ethical:	Exterminated
Juridical:	Exterminated
Aesthetic:	No relevant norms were found.
Economic:	Exterminated
Social:	Exterminated
Informatory:	Exterminated
Historic:	Exterminated
Analytic:	No relevant norms were found.
Psychic:	Exterminated
Biotic:	No relevant norms were found.
Physical:	No relevant norms were found.
Kinematic:	The buyers did move physically in order to visit the wholesale dealer.
Spatial:	Spatial distance between a buyer and a wholesale dealer led to: * shorter distance implied more frequent buyer visits * longer distance implied less frequent buyer visits
Numerical:	No relevant norms were found

The main question of this analysis is: why did the sales process re-design lead to changed ordering behaviour of the buyers? The following answer may be derived from the analysis carried out above.

The reason why the sales process re-design led to a changed ordering behaviour of the buyers was that there was no buyer-salesman social interaction that could give rise to personal and professional social norms, which would establish a trust norm that would found the ethical norms, which would promote a positive buying behaviour. There were no more economic norms that would found psychic norms of motivation for the salesmen, which, in turn, would promote buyers' behaviour positively. The discontinuation of the historic norm implied that the new buyer-IS relation was not historically formed or conditioned, which in this case implied that there were no established norms in the modalities, which impacted the buyers' buying behaviour negatively.

5.   REFLECTIONS AND DISCUSSION OF THE EMPLOYED THEORY

Reflection upon the activities that were performed in this study generates a description of the actual procedure or method that emerged during the investigation. This method had the following main activities.

1. Initial appreciation of the situation, where the investigators became acquainted with the domain of investigation and its circumstances.

2. General description of the situation, including business process model construction of the two situations within the case, identification of motivation for the business process re-design, and identification of the consequences of the business process re-design.

3. Further analysis included:
   a) precision of the main question of investigation,
   b) establishment of focus or boundaries of further investigation; this focus was done by identification of structural changes in actor or entity relations,
   c) further investigation of the focused domain, i.e. the focused actor relation,
   d) multimodal modelling: identification of modal norms within the focused domain in the pre re-design situation,
   e) multimodal modelling: identification of inter-modal relationships within the identified modal norms,
   f) multimodal modelling: identification of modal norm modifications due to process re-design,
   g) derivation of answer for the main question of investigation from the identified modal norm modifications.

In this method description certain issues may be observed. One is that the combination of business process modelling with multimodal modelling was a possible and seemingly fruitful exercise in order to create a well-focused yet multi-perspective description of the situation of investigation.

Second is the problem of focusing or boundary setting for the further elaboration. The problem is of knowing where to set the boundary for further investigation and the risk that answer to the main question of investigation may be outside the defined boundary. This is, however, a general system analysis problem, whose solution is not provided by Multimodal Modelling.[4]

Third is that the norm identifications varied, some focusing on the relation of the two actors or entities present in the investigated domain, and others on an actor as such, though within the focused relation. This variation of focus must be further studied.

Fourth is that the modalities provide a general a priori frame or spectacles that guide the normative inquiry by directing what type of norms it should be attempted to identify. It does not say anything about how these norms should be identified, and thus it may lead to a disregarding of relevant modal norm descriptions. A solution to this can be employment of relevant theories that qualify to each modality. Hence, for example, the psychic modality needs psychological theories, while the aesthetic modality needs aesthetic theory. This leads us to the next issue.

Fifth, identification of modal norms and their interrelations requires an encyclopaedic knowledge of very diverse disciplines. It seems to be a challenging task for a small group of system investigators to possess such knowledge. This in turn may lead to the possibility that multimodal analysis becomes superficial and thus generates a trivial result.[5]

Sixth, identification of inter-modal norm relations was performed mainly on intuitive bases rather than guided systematically by a theory. This questions the reproducibility and usefulness of such an analysis and is an issue that requires further investigation.

Finally, the case study shows that Multimodal Modelling has relevance for system design in general. This is because it focuses on the following questions of system design. What kind of normative change will a proposed system design likely lead to? And then, what consequences will the normative changes lead to?

6. CONCLUSIONS

Multimodal Systems Thinking (MST) is a recent approach to systems design and management in general. It distinguish itself from the more elaborated schools of thoughts, i.e. the so-called Hard Systems Thinking, Soft Systems Thinking, and Critical Systems Thinking, by assuming

---

[4] The issue of boundary setting has been elaborated extensively by Ulrich (1983) in his Critical Systems Heuristics.

[5] [But see chapter 4; MAKE does not depend on encyclopaedic knowledge in the analyst, and Winfield found the modalities easy to grasp by the lay person. Eds.]

different meta-theoretical positions in regard to reality, knowledge, human nature, and particularly the motivational or existential ground-motives (discussed in chapter 13). Even though MST has extensively elaborated theoretical foundations, there are few documented applications within the domain of systems thinking or management. This text has presented an application of the Multimodal Theory, part of MST, in an analysis of a business process re-engineering case. This analysis has provided a multi-perspective description of the inquired situation and also provided an intelligible explanation of that situation. The employment of Multimodal Theory has generated a prototype of a generic method that may be used in future system analyses. This employment has also identified several shortcomings and issues for further study. The conclusion is therefore that MST provides potentially a powerful theoretical foundation for analysis and design of systems, but that this theory needs more research for its operationalization so that it may become an actual tool that supports systems analysis and design.

Chapter 11

AN ASPECTUAL UNDERSTANDING OF THE HUMAN USE OF INFORMATION TECHNOLOGY

Andrew Basden

1. INTRODUCTION

Studies over the past 20 years show the failure rate in information systems (IS) remains high, estimated at around 50% (Lyytinen and Hirschheim, 1987; Whyte and Bytheway, 1996), 60% (Cotterill and Law, 1993; Butterfield and Pendegraft, 1996) and 75% (Gladden, 1982). Much of the high failure rate is due, not to technical failures but to a variety of human factors. Many systems development projects are never completed, or if the IS is completed it is not used, or if used for a time it falls into disuse, or when in use it fails to meet all the users' needs. Even if a system meets the needs of its users, it might have unexpected, detrimental impact, possibly indirectly on other stakeholders or of a long-term nature.

Despite a wealth of literature related to the topic of information systems success and failure the fundamental questions are still far from being answered. In most studies the concepts used are vague or unarticulated and they often neglect the diverse reasons for, and causes of, IS failure (Lyytinen 1988; Lyytinen and Hirschheim 1987).

In this paper, which combines and expands material from Basden (1994, 2001), we attempt to contribute to the debate by throwing light on the nature of usage of information technology (IT) in which success and failure are seen as part of the whole picture. We view usage, success and failure, not from a technical perspective nor an economic perspective, nor even from a socio-technical perspective, but from that of human practice that involves information technology.

2.   USEFULNESS OF IS

To assist our discussion, we will make reference to two cases of IS usage, one large, one small; one a failure, one a success.

2.1   A sizeable failure

We first consider Natalie Mitev's (2001; also see 1996) account of one system that, though delivered 'successfully' in working order by the IS developers, nevertheless was a failure: the SNCF Socrate rail ticketing system. In this description, Mitev's referencing has been removed and the numbers in square brackets have been added, to be explained later:

> "The new ticket proved unacceptable [6] to customers. Public relations [9] failed to prepare the public to such a dramatic change [12]. The inadequate database information [7] on timetable and routes of trains, inaccurate fare information [1], and unavailability [11] of ticket exchange capabilities caused major problems for the SNCF sales force and customers alike. Impossible reservations [8] on some trains, inappropriate prices [13] and wrong train connections [3] led to large [1] queues [2] of irate [6] customers in all [1] major stations. Booked [13] tickets were for non-existent trains [11] whilst other trains ran empty [11], railway unions went on strike [11], and passengers' associations sued SNCF [13]."

This was because:

> "Technical malfunctions [8], political pressure [15], poor management [11], unions and user resistance [15] led to an inadequate [13] and to some extent chaotic [12] implementation. Staff training [9] was inadequate and did not prepare [13] salespeople to face tariff inconsistencies and ticketing problems. The user interface was designed using the airlines logic and was not user-friendly [6]."

This story highlights several issues: poor technical features and training hindering human tasks as a direct consequence for the users, the variety of people that are affected (stakeholders), the diversity of repercussions of the system (in this case, all negative), and various types and degrees of indirect repercussion.

2.2   A small success

A more positive account was provided by the Elsie knowledge based system, which was developed for quantity surveyors in the U.K. (Brandon, Basden, Hamilton and Stockley, 1988).

Elsie was designed to assist surveyors in giving advice to their clients who were at an early (pre-architect) stage in considering the construction of new office developments. It had four modules, a budget module that would help the user in setting an appropriate budget for their client, by inferring the type of building required and drawing on a database of typical costs, a time module that would estimate project timescales, a procurement module that would advise on suitable project management approaches, and a development appraisal module that would assess the long-term financial viability of the project.

Elsie became widely used by quantity surveyors, and this use was studied by Castell, Basden, Erdos, Barrows and Brandon (1992). They found that budget module was the most used, partly because budgeting was a critical task while the others were merely optimising tasks. Every project a surveyor meets is unique, leading to a wide variety of types of situations in which Elsie was used. The users valued the accuracy of the system (which they tested against past known projects), its ability to accept uncertain information, its flexibility in allowing changes to previously entered information, its openness to the user overriding its logic on the basis of local knowledge, the transparent explanation of its reasoning, and that projects could be archived for later recall. As Castell et. al. put it, Elsie "provides a powerful medium for expressing new decisions and clarifying assumptions."

These features led to a number of benefits for the users' tasks. The time to obtain a first estimate reduced from days or weeks to about one hour, and this in turn improved the surveyor's response time to the client.

It also changed what had been a single-stage process (of supplying the client with a detailed estimate based on information received a week earlier) to a two stage process, in which an initial rough estimate was supplied immediately, then the client would meet the surveyor to refine not only the estimate but also the client's requirements with the help the interactive program could give. Castell et. al. (1992) reported:

"To make valid estimates requires a clear picture of the customer's requirements, but it is often the case that such requirements are not clearly known in the initial stages. Customers often change their requirements when they see the implications in terms of costs or building specification. Following the initial stage, the customer and surveyor meet to revise, update and clarify elements of the original estimate, and a new negotiation cycle, based on the outcome, is started. During such cycles, assumptions are revealed and changes in requirements and their consequences are quickly analysed, with the customer present."

As a result,

> "Thus, not only is the process of generating a budget estimate significantly shortened ... but the process enhances the clarity of the customer's requirements. It should be noted, however, that this increases rather than decreases the chances of the customer changing requirements, but such changes can be readily accommodated. ... In addition, direct involvement in the second (revision) stage of the process has meant that the customer has felt more in control of the whole process."

This, together with other task benefits, led to changes in relationship between, and roles of, surveyor and client:

> "The relationship between surveyor and customer has traditionally been one of expert versus novice ... With the kind of shared problem solving behaviour described above, the participants in the negotiation cycle are now likely to be working towards a better articulated, shared goal. Thus the interaction with the computer becomes less differentiated in terms of expert versus novice and more co-operative in terms of reaching the common goal of an acceptable budget estimate and building specification. The change in relationship is towards empowerment of the customer ..."

This clearly displays the network of interrelated factors in the use of such software in professional situations of decision making and advice giving. The revealing of assumptions was particularly important in improving the quality of open dialogue between the parties. As Basden (1994) discusses, the impacts are at three levels: features of the software use (e.g. flexibility), which enabled change in tasks (e.g. single-stage to two stages), which in turn led to change in role and relationship.

2.3   Issues in usage

We have so far identified several issues in regard to usage and usefulness that complicate our understanding of it, and thus our ability to evaluate, predict or design information systems:

♦ Multiple stakeholders - Not only is the primary user affected, but many others too. In the case of Elsie it was mainly the primary user and their client who were affected, though the user's organisation may also be seen as a stakeholder. In the case of Socrate, the stakeholders mentioned by Mitev were: management, railway unions, user, railway staff, salespeople, passengers ("queues of irate customers"), 'the public', trains that ran empty (note a non-human stakeholder), passengers' associations, and SNCF itself.

♦ Diversity of impacts - In the case of Elsie, accuracy and speed of making estimates were enhanced, client requirements were clarified, and the relationship between, and roles of, surveyor and client changed. In the case of Socrate, there were: technical malfunctions, political pressure, unions and user resistance, inadequate and chaotic implementation, tariff inconsistencies, ticketing problems and general unacceptability, inaccurate fare information, and unavailability of ticket exchange capabilities, impossible reservations, inappropriate prices, wrong train connections, large queues of irate customers, tickets for non-existent trains, trains running empty, strikes, and litigation.

♦ Indirect impact - the direct impact on people might cause a chain of repercussions that is difficult to predict or control. In the case of Elsie, faster initial estimates and the ability to explore possibilities stimulated a challenging of assumptions, which led in turn to the client changing their requirements in ways that were considered helpful, and this changed the relationship between surveyor and client. In the case of Socrate, the picture is much more complex, with technical problems and poor system design, together with poor training of staff leading first to direct problems for the users, such as errors in reservations. This had repercussions for those people the users had to deal with, the customers. It also had repercussions on the running of trains (some empty, others non-existent). These problems, in turn, had wider repercussions such as strikes and passenger associations suing SNCF. Unfortunately, neither study was designed to consider yet wider repercussions, such as ecological, which occur when a system becomes very widely used. But it is not unreasonable to ask what ecological impacts might occur. For example, had Socrate's problems not been remedied, there might have been modal switch from rail to road, with consequent increase in climate change emissions. In the case of Elsie, the possible ecological change is more subtle. At the time, the tax system favoured wasteful new-build over relatively ecological refurbishment. Since Elsie was designed only for new-build, its widespread use might have exacerbated this anomaly.

♦ Unexpected impact - using the computer system might bring the benefits that were originally identified as desirable, but it can also bring detriment in other areas that were not considered, constraining other tasks the users must complete and perhaps encouraging deficient (e.g. lazy) ways of working. In the case of Socrate, most of negative repercussions were completely unexpected. But unexpected benefits can also materialise, such as in the case of Elsie, which had been designed mainly to allow budget setting to be carried out by less expensive personnel. That it facilitated open discourse and changed the surveyor's way of working was a surprise to us. The change in client-surveyor relationships was even more unexpected, especially as it was a change at a different level, that of social role rather than task.

♦ Normative content - Given the accounts above, it is reasonable to see Elsie as a success and Socrate as failure. But on what basis is it valid to see them thus? Why should "impossible reservations" and strikes be negative while "flexibility" and shortening a process be positive? Sometimes one stakeholder benefits while others are negatively affected. It is not uncommon for there to be benefit in the short term but detriment in the longer term, or for there to be visible benefit but detriment that is less visible. To answer this, we need a framework for understanding usage that recognises normativity as such, without reducing it to either mere description or opinion. It must recognise a diverse normativity, which extends to (e.g. non-human) stakeholders that have no voice (Basden and Wood-Harper, 2005), and that impact could be positive in one way, to one stakeholder, while negative in another way, to another stakeholder.

3.   A PHILOSOPHICAL APPROACH TO USAGE

This paper addresses these questions by means of the positive component of the philosophy of Dooyeweerd (1955). Based on Meaning and normativity, this philosophy can provide interdisciplinary frameworks for understanding such topics as we wish to discuss here: usage, success and failure in IS. We make a very outline examination of a subset of Dooyeweerd's philosophy here and suggest how it might be applied to understanding usage. This will throw light on the diversity of repercussions and impact of use and give us a framework by which we might differentiate success from failure, especially in cases where one stakeholder might benefit and another experiences detrimental impact.

3.1   Aspects of reality

As an integral part of his philosophy, Dooyeweerd (1955) proposed that reality as we experience it has a limited number of aspects (otherwise called dimensions, modalities or spheres) that are irreducible to each other, and each aspect has laws that pertain when entities (people, sheep, roses, pebbles, etc.) function in that aspect. In a full version of his philosophy, this is tied up with Meaning and Time and God, but we do not explore those issues here. He proposed an initial set of fifteen aspects, each of which has a kernel meaning (in brackets):

1. Quantitative (amount)
2. Spatial (continuous extension)
3. Kinematic (movement)
4. Physical (energy)
5. Biotic (life functions)
6. Sensitive (sense, emotion)
7. Analytical (distinction)
8. Historical-technological-cultural (formative power)
9. Lingual (symbolic meaning)
10. Social (social intercourse)
11. Economic (frugality)
13. Aesthetic (harmony)
13. Juridical (what is due)
14. Ethical (self-giving love)
15. Pistic-credal (faith, vision, commitment)

(This set of fifteen is merely a proposal that he expected to be developed, but it has so far proved remarkably robust. In this paper we accept it, but recognise that adjustments might be required later. Also, in this paper we will name the 8th aspect 'Formative', as being more succinct and also more descriptive of its true meaning than 'Historical-technological-cultural'.)

3.2   Functioning

The aspects are law-spheres that enable meaningful functioning which has repercussions. Aspectual functioning is a complex issue, involving an understanding of subjects and objects that is non-Cartesian: Dooyeweerd held that being a subject or object is not based on what we are (human thinking 'I' versus non-human 'it') but on how we relate to the laws in each aspect. A subject is so in that it/he/she is subject to laws of an aspect and makes a response to those laws. Thus, for example, I am subject to the laws of the lingual aspect as I write this, and so is the reader when reading. We may be described as lingual subjects. A rock is subject to the laws of the physical aspect; it may be described as physical subject (not 'physical object', as we usually do under influence from Descartes). An object is so in that it/he/she is involved in another entity's subject-functioning. So, for example, the ink on paper is a lingual object, because it is involved in my lingual subject-functioning of writing. An entity may be subject in one aspect, object in another: the ink is both lingual object and physical subject.

The earlier aspects are determinative in character while the later aspects are normative. Experience of freedom comes not from what we are ("I am free and active; the rock and ink are determined and passive") but from the extent to which the aspects in which we function as subject are determinative or allow freedom. I am free in how I express myself in writing because the lingual aspect offers me freedom. But the rock and the ink, as physical subjects, are physically determined - and so am I - because the physical aspect is determinative. Human beings are subject to the laws of all aspects, but animals, plants and non-living things are subject to laws of aspects up to the sensitive, biotic and physical respectively.

It is as we subject ourselves to the laws of aspects that behaviour, functioning, activity occurs; the aspects constitute a framework which makes possible all that occurs. Note Dooyeweerd's novel view of law, not only as constraint but also as enabler of meaningful activity; without aspectual laws there would be no activity. While determinative aspectual law provides only one possible outcome for each subject-response to it, normative law allows and enables freedom and defines what types of activity are meaningful and likely to be of benefit.

3.3   Multi-aspectual human functioning

Our activity in life is in fact a being-subject to the laws of every aspect, as depicted in Fig. 1. Thus human activity is complex not only in what happens but also in the multi-aspectual nature of what happens. And this includes our using IT. For example, as I use my word processor to write this, I am functioning lingually, but I am also functioning formatively as I structure my thoughts and sentences, analytically, as I decide what to say and what to leave out, sensitively, as I see the screen and my fingers press the keys, biotically, as I breathe and digest my food, and so on. I also function socially, in that I see the reader as a person with whom to relate rather than a mere recipient of information, economically, since I must keep to a certain word limit, and so on. Kalsbeek (1975) discusses how all aspects (even the pistic - our vision of what we are) play an important part in manned space flight.

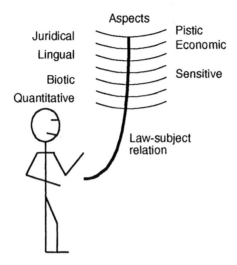

Figure 1. Functioning in all aspects: the law-subject relationship

3.4 Impact of using IT

As I function (e.g. use IT), there are repercussions in all, or most, aspects. Each aspect yields a different type of repercussion. Such a diversity of aspectual repercussions helps to account for the diversity of types of impact that using IT brings about. See Fig. 2.

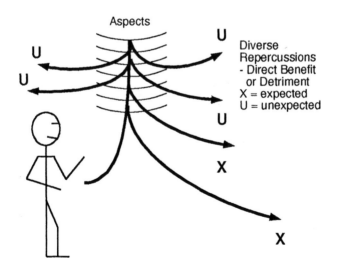

Figure 2. Diversity of aspectual repercussions of human activity

For example, if I create a web page, lingual repercussions include the reader understanding what I mean, or not, and social repercussions include that if my writing is blunt, the reader is less likely to be friendly towards me and thus less likely to accept its content.

As discussed by Winfield and Basden in chapter 4, when we function, we are unaware of many of the aspects. In planning or evaluating our usage of IS, we might be aware of some of them, and in these we might expect repercussions (X), but others we overlook and repercussions in those are unexpected (U). Repercussions in those aspects continue to occur even though we might not be aware of them.

The aspectual repercussions of our functioning impact on other entities around us, as objects. However, those objects of our functioning are also subjects in their own right and their subject-functioning changes as a result of our impact on them - which in turn has its own repercussions; see Fig. 3. This is how Dooyeweerd would account for indirect impact of using IS, and all involved are what might be called stakeholders. Note that the stakeholders might be human or non-human, so this gives us a basis for addressing impacts of, for example the environment (see the biotic and physical subject-functioning of the tree in Fig. 3).

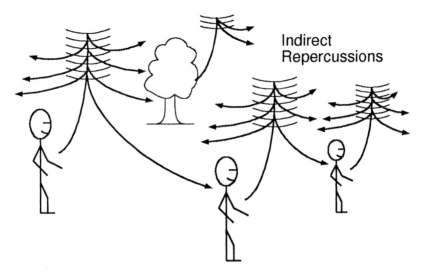

Figure 3. Indirect aspectual repercussions

This is a simplified picture. But it is able to provide an account for the first four of the issues of usage we identified earlier:

- ◆ Diversity of stakeholders occurs not only because several entities are objects in our subject-functioning in an aspect (e.g. several people reading this), but also different stakeholders may be defined by reference to different aspects; for example, readers are lingual stakeholders of this book, the publisher is an economic stakeholder.
- ◆ Diversity of repercussion arises from diversity of aspects in which functioning occurs even in a single action - in principle every aspect.
- ◆ Indirect repercussions occur because of the impact on other entities shown in Fig. 3.
- ◆ Unexpected repercussions occur because we have taken too little account of functioning in certain aspects - often the later ones.

With some aspects the effects can be almost immediate but with others, such as the pistic, the effects are long term, so unexpected repercussions take a long time to appear and often it is difficult to link repercussions retrospectively with the functioning that led to them.

3.5 Normativity

Normativity gives us some freedom and also some responsibility. Each normative aspect conveys different norms such as (for example):

- ◆ Lingual aspect: Utterances should be well-formed and should not deceive.
- ◆ Social aspect: Be friendly and courteous rather than rude.
- ◆ Economic aspect: Be frugal rather than wasteful.
- ◆ Juridical aspect: Give each person, animal, group, etc. what is their due.
- ◆ Ethical aspect: Go beyond what is due, even to the point of self-sacrifice. Be self-giving rather than selfish.
- ◆ Pistic aspect: Be committed, be courageous, have vision, take risks, but our vision and commitment should be centred on the true Deity rather than any substitute for Deity.

Central to Dooyeweerd's proposal is that if we function in line with the laws of aspects (which we may call positive functioning), then things will, in general, go well, but if we go against the aspectual laws (negative functioning) then things will not go well. This is the basis for understanding the difference between benefit and detriment. This includes our use of IT. For example, if we are selfish in our use then we upset people and will no longer have their best cooperation, and indeed the whole atmosphere in our organisation might become damaged.

Another tenet of Dooyeweerd's philosophy is that there is no inherent disharmony among the aspects, and it is both possible and desirable to function in line with the laws of all aspects. Therefore, if either we create an imbalance between the aspects, ignoring some and putting undue emphasis on others, or we go against the laws of one or more aspects, then our functioning will be negative in various ways, and so will the repercussions. This accounts for why, for example, a user who is very logical (functioning in the analytic aspect), produces excellent results in their use of software, may also be very selfish (ethical aspect) and be so difficult to work with that the team becomes dysfunctional.

(The laws of each of) the aspects are irreducible to each other. For instance, laws of the ethical aspect cannot be derived from those of the social or economic, and vice versa. What this implies is that if we want to maintain balance among the aspects we must attend to each aspect explicitly, rather than assuming that e.g. ethics or justice will be taken care of.

3.6   Success and failure in using IT

This view can help us define or at least understand success and benefits and differentiate them from failure and detriment. We can portray the functioning or repercussions in the aspects graphically as a 'Christmas tree', a double bar chart in which the degree of positive functioning in each aspect is shown as bars to the right and negative functioning as bars to the left. Fig. 4a shows a general example. Fig. 4b, where all bars show negative, is the aspectual profile of the SNCF Socrate system as described earlier in Mitev's account, in which the bracketed numbers refer to the aspects. The lengths of bars express the number of times each aspect is found in the account.

Though a crude device, it can give us an overview of the aspectual profile of any particular usage of IT. Groups of negative bars in the tree point to areas of failure while groups of positive bars indicate areas of success. Notice that there may be positive functioning in some aspects an at the same time negative functioning in others. It is even common to find both positive and negative functioning in the same aspect.

It should be borne in mind that Mitev's account, though it clearly displays a wide variety of aspects, was not written with Dooyeweerdian aspects in mind, nor even an aspectual analysis. Our reference to her account not only illustrates the diversity of problems, but it also illustrates how an aspectual analysis might be undertaken of existing texts.

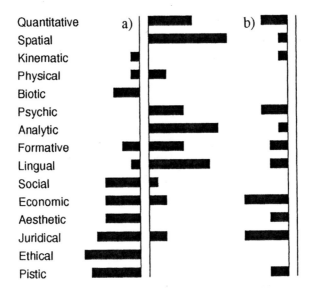

Quantitative
Spatial
Kinematic
Physical
Biotic
Psychic
Analytic
Formative
Lingual
Social
Economic
Aesthetic
Juridical
Ethical
Pistic

Figure 4. 'Christmas Trees' showing profile of aspectual functioning
(a) Hypothetical general case (b) SNCF Socrate system

4. DISCUSSION

4.1 Use of the aspectual analysis

Success and failure of IS is not a simple issue. In many cases, we cannot say whether an IS is unequivocally one or the other. The kind of aspectual analysis described here can throw light on the issue. It provides a framework and a graphical tool by which the aspects of success and those of failure may be separated out, so that the main areas of problem may be identified and appropriate action be focused effectively.

The 'Christmas trees' of aspectual functioning described above is a multi-purpose graphical tool but it is just a start. It presents a danger of too quantitative an approach in which we give too much attention to the pseudo-quantities designated by the lengths of the bars. Instead, we should look for patterns, such as clusters of either positive or negative functioning, and use these to tell us where further and deeper analysis is needed.

The analyses stimulated by the above considerations can be taken further by using SSM as discussed in chapters 3 and 5, or be carried out using the MAKE methodology described in chapter 4. This is a method for analysing a domain of ill-structured knowledge in a manner that stimulates consideration of every aspect by those involved in the situation.

It has proved to be very usable and easy to learn, and highly effective in obtaining wide aspectual coverage.

4.2   The validity of the aspects

Though Dooyeweerd made an ontological claim for aspects as such, he recognised that both his own proposal for what they are, and also our collective theoretical knowledge of their laws gleaned through science, will always be partial and flawed. Therefore, as we employ a suite of aspects for tasks like IT evaluation, we should always be open to the possibility that our understanding of them stands in need of refinement.

However, does this mean that we must be dominated by skepticism? Dooyeweerd would answer "No!" for two reasons. One is that, though we can never fully understand the kernels of the aspects by means of *theoretical* thinking, they may be *intuitively* grasped. We can never explain fully what justice is, for example, but we intuitively know, understand and recognise it, even if our intuition sometimes varies. Winfield's (2000) findings support this: after a suitable short period of learning, the aspects were understood by lay clients sufficiently well as to make significant progress in MAKE.

The other is that, though Kant claimed we can never know the 'Ding an Sich' so that thinkers since Kant have presupposed what Tarnas (1991) calls the radical illegibility of the world, Dooyeweerd claimed that Created Reality tends to reveal itself to us rather than hide itself. That is, though our (analytical etc.) functioning is flawed, it tends to be in line with the aspects as they really pertain.

However we still have to justify why Dooyeweerd's suite of aspects should be better than anybody else's. Some (Hart, 1984; De Raadt, 1997b) question the precise number of aspects though they do not propose any major modifications to Dooyeweerd's suite. Others, coming from very different backgrounds, have proposed suites of aspects without any reference to Dooyeweerd. Checkland suggested Five E's, Maslow gave us his famous hierarchy of needs, Bunge (1979) spoke of emergent systems. Dooyeweerd's suite seems superior to most proposals for the following reasons:

1. Coverage. Dooyeweerd's has wider coverage; most other suites are found to be subsets of Dooyeweerd's. For example Checkland's (1991:A25) five E's map quite directly to five of the aspects as shown in Table 1 in Mirijamdotter and Bergvall-Kåreborn's chapter 5.

2. Dooyeweerd's theory of modal aspects is part of a larger, coherent philosophy, not just a suite that has been proffered.

♦ Dooyeweerd related his theory primarily to the everyday lifeworld more than to the products of theoretical thought.

♦ He tackled the problem of how diversity is to remain coherent rather than ending up as mere fragmentation, by positing not only irreducibility between the aspects, but also relatedness of specific kinds (referred to in chapter 2).

♦ Dooyeweerd has made philosophical proposals about what constitutes a distinct aspect. For example, that it lacks antinomies, which arise from conflation of two aspects.

♦ Being based on Meaning rather than Being, his suite is particularly amenable to handling the purpose and meaning we encounter in everyday use of IT.

3. Scrutiny. While all suites must emerge from everyday experience (Dooyeweerd argues for the importance of the latter), Dooyeweerd's suite has also been subjected to philosphical, teleological and historical scrutiny.

♦ The philosophical scrutiny has involved seeking antinomies within the aspects, and setting the aspects within a framework of Meaning.

♦ The teleological scrutiny has involved discussing the role of each aspect in the total spectrum of Meaning.

♦ The historical scrutiny has involved a survey of 2,500 years of Western thinking to detect the aspects that thinkers have believed to be important, and how they treated them.

4. Because of the latter, Dooyeweerd's suite is likely to have a cross-cultural and trans-contextual applicability that most other suites lack.

5. Personal Qualities.

♦ Dooyeweerd reflected at length on the aspects of the lifeworld, with no intellectual axe to grind within the conventional debates. Checkland (1981), by contrast, explicitly takes an interpretivist stance, and Bunge (1979) admits that his choice of system levels (aspects) was determined by his distaste for what he considered mystical.

♦ Dooyeweerd's was self-critical in the Habermasian sense, aware that presuppositions underlie all theoretical thinking, including his own. It is not clear to what extent the proposers of other suites are or were.

So we are justified in adopting his suite as a starting point for IT evaluation, even though we may refine it sensitively.

5.   CONCLUSION

We have presented an approach to understanding the usage, success and failure of information technology that both accounts for troublesome issues such as indirect and unexpected impact, and is able to differentiate success from failure, even in complex cases in which benefit and detriment are mixed together and might accrue variously to different stakeholders. In this way, we have been able to bring success and failure together into a single framework of understanding.

The framework is based on Dooyeweerd's notion of irreducible aspects of meaning, which enable functioning, ensure repercussions and define norms. We have suggested a simple graphical device that could be useful during analysis, and shown how it may be used in practice. We have also discussed the validity of Dooyeweerd's suite of aspects. It is interesting that this aspectual approach is currently being explored to understand not only IT usage, but also environmental sustainability (Lombardi, 2001; Brandon and Lombardi, 2005).

Part IV: Directional Perspectives

Chapter 12

THE IDEA OF A SYSTEMS ETHICS

Sytse Strijbos

1.   INTRODUCTION

In chapter 6 we discussed the systems character of modern technology. Modern technology began with machine technologies, to be sure, but its influence is much more pervasive today. This fact that modern technology embraces so much more than separated things that stand apart from us for our appropriation and use is fraught with implications for human action and for the ethics of technology. Individual actions with technological things are executed in human surroundings created by technology. If, for example, we use the car to get to work, we participate in a collective system of traffic. Or if we take a shower, we use collective utilities for the purification and distribution of water. Both these examples show that it is not so much machines as such but systems that have assumed primary importance. And not only things but also people have grown to be dependent in their functioning on the technical world of organizational systems (Ropohl, 1979). Everywhere today - whether in industry, tourism or commerce, in health care or social work, in politics, the arts or anything else - we are confronted with collective systems. These are givens, and they determine to a high degree how individuals act. Ellul (1964:xxviii) has this in mind when he states "that there is a collective sociological reality, which is independent of the individual ... individual decisions are always made within the framework of this sociological reality, itself pre-existent and more or less determinative."

Naturally, strong behavioral influences arise from the fact that ordinary human life unfolds today within an ever expanding complex of regulations, organizational connections and systems. Yet to say no more than that would be to speak a half truth. The sociological reality of collective systems is not just a given (datum) that precedes action and creates and defines a certain space for it; it is at the same time a task (dandum) that challenges us with new dimensions of human responsibility and a mandate for ethical reflection (Jonas, 1974:3-8; Jonas, 1979 and May and Hoffmann, 1991). There has been an increase in the scale of human action that brings with it an expansion of the terrain of ethics. For example, as recent decades have shown, ethical reflection regarding health care may no longer be focused exclusively on problems in the doctor-patient relationship. Ethics today must also take into account the fact that the individual physician is gradually being superseded by a complex medical enterprise, which is in turn just part of a complex system of provisions for which government authorities bear a significant share of the responsibility (Strijbos, 1994).

In short, the technological society confronts us in many fields with ethical questions regarding the normativity of human existence as we live it concretely today in a self-constructed world, namely the world of the various collective systems which people have developed and which they control themselves. Such questions belong to systems ethics, which may thus be defined as a specific field of ethical reflection. Just as there are various schools of individual ethics, so too there are various possibilities for the development of a normative systems ethics. Each expresses a different approach to modern, technology-dominated society. A bit schematically we will distinguish three approaches. We shall look first at the 'ethics of adaptation and control' proposed for modern society by Laszlo. Then we shall consider its counter pole, an 'ethics of liberation', as in Ellul. Against the background of these conflicting positions, we argue finally for an 'ethics of disclosure' for the continuing development of the technological society.

2.   ETHICS OF ADAPTATION AND CONTROL

In the western world during recent decades, the optimistic belief in technological and economic progress which inspired many thinkers of the last century has suffered a serious shock. The word 'progress' has lost its charm, yes, has even acquired a negative connotation instead (Goudzwaard, 1997). The compass to which the industrial world sets its course seems now to have proven unreliable. The great question has thus become: How are we to proceed from here?

It is this climate of thought in which the systems philosopher Laszlo desires to reorient the technological-industrial society. Equipped with the modern powers of science and technology, he says, modern capitalism was able to realize unequalled growth and economic productivity. "Its values were materialistically oriented: *the good* is a large production per capita, and *the better* a still larger production." (Laszlo, 1972b:102) In the meanwhile, Laszlo observes, it has become clear that western society, with these materialistic values, has run aground. In the first place, production per capita cannot be increased indefinitely. Moreover, the earth is not rich enough for the entire world to attain the standard of living to which western society has grown accustomed. There are thus limits to growth. And this means that there is an urgent need for a new guiding ethic and a revised definition of what we are to understand by 'progress'.

"We are at cross-roads ... Thus what we need today is a new morality - an ethos which does not center on individual good and individual value alone, but on the adaptation of mankind, as a global system, to its new environment." (Laszlo, 1972b:281)
"Progress cannot lie in more and bigger. Progress must be redefined and that means a new system of values." (Laszlo, 1972a:103)

Finding a new pattern of values Laszlo thus sees as the most urgent problem of our time in need of a solution. Namely, cultures are 'value-guided systems'. A culture therefore has no future if it lacks a unifying, guiding system of values. Or, to say the same thing in systems terminology: every culture must, as a dynamic, open system, be stabilized by 'adaptive self-stabilization' or 'self-stabilizing negative feedback'. The new system of values which our technological society has such great need of would in fact fulfill such a systems-cybernetic principle.

A new social-cultural ethic is thus necessary, but is it also possible? It does not elude Laszlo that insofar as the western world is concerned, we have arrived here at a delicate problem. Namely, there is widespread scepticism regarding questions of normativity. Not only does one find that people uphold entirely different norms and values; it is also widely held that there is no longer any basis for regaining a certain consensus. Norms and values are regarded as no more than personal preferences. And the consequence is then that every attempt to arrive at an overarching ethic is doomed beforehand to failure. A cultural community that is governed by a spirit of scepticism, moral relativism and pluralism is bereft of every possibility for an objective grounding of normativity. Norms are private in character and can therefore offer no direction to public life (Laszlo, 1972a:104, 1972b:270-72).

There is something remarkable about Laszlo's description of modern industrial society. For on the one hand he maintains that technological-industrial society has been ruled, at least until recently, by a sacred belief in progress and the materialist values that go with it, while on the other hand he sees that there is a spirit of scepticism and moral relativism in the West. What is the connection between the two? Has one fathomed the situation of modern western culture deeply enough if one observes that, following the failure of the ideal of progress, a spiritual and moral vacuum had to set in? Or is there more at hand? Is there more to be said about the crisis of meaning and the cultural malaise about which Laszlo is so rightly concerned? Are there no more fundamental causes than failure of the belief 'that the sky was the limit'? (Laszlo 1972a:103) Indeed, there are, and Laszlo knows it. The belief in progress does not stand alone but is according to Laszlo an expression of a typically western total view of reality, a view that is, so he maintains, entirely alien to traditional eastern thought. The anthropocentric attitude of western man whereby on the basis of a utilitarian ethic people exploit limitlessly for their own purposes the world in which they find themselves Laszlo (1972b:283) attributes to Christianity and the mechanistic Newtonian world picture:

> "While in the Orient, philosophy and religion fused in the appreciation of nature as 'self-thusness' (*tzu-jan* in Chinese and *ji-sen* in Japanese); as something *for-itself*, rather than *for-us*, in the West there arose out of Christianity a Puritan ethic of hard work for human well-being and the glory of God, which was premised on using nature for human ends. With the contemporaneous rise of science and the scientific attitude typical of the mechanistic Newtonian world-view, utility was conceived as the finest goals of knowledge. Bacon wished to 'subdue and overcome the miseries of humanity' by discovering in the womb of nature 'secrets of excellent use'. Hereby man's attitude to nature changed from one of *reverence* to that of *exploitation*. Whereas Eastern man continued, at least until our day, to revere nature for what it is in itself, Western science and technology appreciated it mainly for its use."

In response to this lengthy citation, one might pose various critical questions. Is it correct, for example, to posit a direct connection between western anthropocentrism and Christianity? Moreover, is it actually the case that traditional Oriental religion and philosophy evince little affinity to mechanistic scientific thought, as Laszlo suggests? Suffice it to say that the British theologian Lesslie Newbigin and others maintain precisely the opposite. Traditional Chinese and Japanese religions can be united without any particular difficulty with the materialist scientific worldview, while there has always been a tension between it and the Christian religion. (Notice that the reference is to a tension between Christian religion and the scientific worldview, not to a contradiction between Christian religion and science.) "The reason is clear. The Eastern religions do not understand the world in terms of purpose ... The Bible, on the other hand, is dominated by the idea of divine purpose." (Newbigin, 1986:40)

From the preceding it follows that there can be no breaking through the ruling pattern of norms and values and notions of progress without a renewal of a total view of reality. The times are ripe for it, too, Laszlo believes. Western anthropocentrism, and scepticism too, can be transcended by attuning the scientific world picture to the latest scientific insights of the systems sciences. This means to Laszlo that it is not the Cartesian subject that is the measure and centre of reality. The starting point must be located rather in the complex hierarchy of systems, developed in the course of evolution, in which this subject is incorporated. According to 'the systems view of nature and man', the highest value does not lie in the individual subject, no, "the norm, ultimately, is given by the highest supra-system: the hierarchy." (Laszlo, 1972b:284)

Now, what does this mean for our pluralistic, technological society? What new social ethics offers us the possibility of charting a safe course towards the future? The conclusion to be drawn from what has just been said seems obvious. If the systems in which human life transpires are a hard fact, then this means that man will have to obey all their laws. In that case these systems and their goals form an objective basis for a new pattern of norms and values. The technological progress of the last hundred years, which fill people with such pride, turns out to have been an enormous blunder in the process of social evolution. It has put the future of humanity at stake. This development can no longer be reversed. People will therefore have to learn again how to adapt themselves to the natural order and to respect it instead of myopically exploiting it. "We can do this, if at all, only through the rediscovery of classical Hellenic and Oriental modes of thought as the inspiration for a new-age ethos which reveres the natural order, rather than *exploits* it." (Laszlo, 1972b:287) The social ethos for the technological global society of the future is: "*reverence for natural systems.*"

What is striking in all this is that Laszlo speaks of a 'natural' order and of 'natural' systems in such a way as to include in them the world that man has created artificially through technology. Within the framework of his evolutionary philosophy, the boundary between nature and culture blurs. This is clear not only from the fact that technological society is assigned a place in the realm of nature but also from the fact that the realm of nature is interpreted by Laszlo with the help of concepts borrowed from technology. Implicit therefore in the reverence for the natural order which Laszlo commends to us as the social ethos for the future is reverence for the man-made world of organizational-technical systems. As it turns out, then, the systems ethics propagated by Laszlo is in fact an ethics of adaptation to this world. The stability and survival of the technological world form the all-controlling goal to which everything must adapt. So too, in the end, must ethics.

With 'ethics of adaptation and control' I do not mean to imply that Laszlo holds a conservative view of society, i.e. an adaptation to the status quo. Such conservatism has often been attributed to certain social systems theories (see Luhmann, 1970:114). Even the theorists themselves have been characterized as overly conservative (Collins, 1975). From Laszlo's work, however, it is completely clear that he does not regard the system as a fixed structure. With an 'ethics of adaptation and control' is meant here that the system is regarded as the normative horizon for human acting. Norms, or what Laszlo (1995) calls 'the sustainability parameters' of 'natural systems', are than in fact derived from the (dynamic) system, rather then that norms act as guiding principles pointing us the way in (re)designing the systems in which our lives are embedded.

Although we share Laszlo's (1977, 2001, 2003) deep concern about the future of our societies, we believe that we need another perspective on reality and on man's place in it. Before going into it, let us first listen to the French social critic of the technological society, Jacques Ellul.

3.   ETHICS OF LIBERATION

There can be no doubt that Ellul will be very critical of Laszlo's conception of systems ethics. He will see in it a confirmation of the autonomy of technology. For Ellul the problem of our age is namely located on an entirely different level than Laszlo seems to think. While Laszlo opposes the anthropocentrism and exaggerated emancipatory drive of western man, Ellul locates the problem in the autonomy he ascribes to the technological system. Hence he will resist an ethic which seems to legitimate this independence. In opposition to the coercion of the technological system, Ellul (1986) advocates the freedom of the subject and adopts an 'ethics of freedom'.

How are we to conceive of such an ethics? What is meant by the coercion of the technological system? Is Ellul not too negative in his assessment of technology, for has technology in fact not only enlarged human freedom? At least, that is the usual picture of technology. Toffler emphasized again and again, for example, in *The Third Wave* (1980), that in the third-wave society technology creates more freedom of choice for people. The new form of society that is set in motion by the the Third Wave breaks through the mass character of industrial society. The individual is restored to the center. People are free to choose their own lifestyles, to decide their individual behaviors for themselves, and to pursue their own patterns of consumption. Again, the question is, does ongoing technological development not bring ever more freedom? Is Ellul's critique of the technological society not inspired by a romantic view of societies that still live close to nature?

Yet such considerations do not form the background, as it turns out, of Ellul's critique of modern technology. He recognizes that in a traditional society in which human life and community are still largely imbedded in the realm of nature, people are anything but free. For people in such circumstances are constantly threatened by numerous dangers from their surroundings - by the dangers of disease, famine, cold or heat, lack of water to drink, and so forth. People are the prisoners of their natural surroundings, which condition every aspect of their existence. "Beyond the shadow of doubt man enclosed in nature is in no way free ... He is locked into a struggle for his survival that was surely no freedom." (Ellul, 1980b:273) Moreover, so Ellul continues, everything that science teaches about nature has the character of 'necessity'. Even the concept of 'chance' speaks not of freedom but of an order of necessity. Freedom can be said to exist only when nature is constrained, that is, when the laws of nature are used to bring about something new, a new artificial world. In Ellul (1980b:274), therefore, the concept of 'freedom' has something to do with what is artificial, including technology:

"Man is free only in his own invention, when he builds an environment for himself (society) and gives himself tools for it (knowledge, religion, art, and technique). The natural realm is the world of necessity; the artificial realm is the expression of his freedom."

However, although Ellul regards technology as an expression of human freedom and the way to his freedom, this is for him decidedly not a metaphysical, immutable and eternal truth. In the process of historical development a reversal of sorts can occur: what served at first as an instrument for the attainment of freedom can become a condition for a new kind of slavery. It is precisely that that happened with technology during the development of industrial society. Technology grew into an unbridled, autonomous power that subjected everything to itself. The artificial won a complete victory over the natural. Yet when people create a completely artificial environment for themselves, they end up in an uninhabitable world. Their negation of nature leads to a reversal: people come to be ruled by the artificial world, as it comes to be independent of them. When the artificial world loses its connection with the natural world, then the human freedom that was attained in the artificial world evaporates. People become prisoners of the technological system they themselves have designed. "In effect, this is what happens to us globally, collectively. In return for having made a totally artificial universe, having eliminated the natural constraints, we have placed ourselves in a world where freedom is no longer possible at all." (Ellul, 1980b:278)

When Ellul so forthrightly asserts that human freedom has again been lost, then this could easily lead to misunderstandings. He does not mean to assert, namely, that people are mechanized and conditioned as if they had become some sort of robotlike creatures. Within the system people are still assuredly able to make choices. Technology presents a rich scale of options, richer than ever before. People can choose - to take an arbitrary and relatively harmless example - between telephones of a great variety of colors, models, and capacities.

Naturally, we are existentially more involved if the choices confronting us involve not things but matters of life and death. Modern medical technology confronts us with such choices. Ethical questions concerning abortion, euthanasia, in vitro fertilization, and much more are the subjects of animated discussion in the West today, in the medical world to be sure, but also especially amongst the broader public. Concretely at issue are the possibilities medical technology offers of prolonging or ending life artificially, or, moreover, of allowing someone to die painlessly, yet with the consequence that one is not aware of one's dying. Now, does this sort of new medical-technological power increase human freedom, asks Ellul? Does it not much rather deprive one so dying of one of the most important moments in life?

The matter is one of choice, of course. Ellul does not deny this. What concerns him is that the options are always already defined beforehand by the technological environment. No choices are available to people except those given with the technological system. People's actions are determined by the technological system. At best there may be no more than a certain tolerance within its strict boundaries. Freedom to adopt a position opposing the system is lacking. Therefore Ellul (1980a:325) regards it as a myth to speak of increased freedom of choice as a result of technology.

"Man can choose ... But his choices will always bear upon secondary elements and never on the overall phenomenon. Man can choose, but in a system of options established by the technological process. He can direct, but in terms of the technological given ... The human being who thinks and acts today is not situated as an independent subject with respect to a technological object. He is inside the technological system, he is himself modified by the technological factor."

But is there according to Ellul then no escaping the system? Or is there for human beings in the system still and again a way that conveys towards freedom? A return to the past or to nature is excluded. We have no way of undoing the past that lies behind us. The technological system is a given with which we shall have to learn to live, as we once learned to live with nature. The question is therefore: Can we overcome determination by the technological environment as we once overcame determination by nature? Ellul's (1980b:281) answer is in the affirmative. With the first level of the artificial - technology - behind us, we now confront the challenge of developing a second level of the artificial. And "the second level of the artificial is ethical."

It is our task to develop a new ethics - not an ethics based on the principle "freedom through power" but an ethics based on "freedom through 'non-power.'" Here 'non-power' does not mean 'powerlessness' but voluntary limitation of the power of control. "Non-power does not mean giving something up, but choosing not to do something, being capable of doing something and deciding against it." (Ellul, 1980c:245) The second level of the artificial means not that technology liberates us but that we liberate ourselves from technology. As human beings once claimed their freedom from nature, so we today must reclaim our freedom from the technological system. Emancipation from technology as a power determining us can be achieved by demythologizing and desacralizing technology, as was once done with nature.

4.   ETHICS OF DISCLOSURE

In the foregoing sections it has been pointed out that Laszlo and Ellul offer widely divergent perspectives to the technological society and that the differences are concentrated in their evaluations of the position of the human subject with respect to the technological system. In conclusion, I should now like to say that where Laszlo calls for reverence for the system, Ellul advocates precisely the opposite: against the unholy system he proclaims once again the subject, who refuses to adapt to the coercive order of the system but is prepared instead to fight for human liberty. It is not the destruction of the system that he desires, "it is simply an attempt to deal with a new environment which we do not know very well. It is a matter of reaffirming ourselves as subjects." (Ellul, 1980c:247)

Neither Laszlo's nor Ellul's efforts to establish an ethics for the technological society seem satisfying to me. Their opposed conceptions evince a fundamental tension between subject and system. Its cause is to be found in the Western subjectivism that dominates both their schemes. Granted, Laszlo endeavours to relocate the self-defining subject in the world of which this subject is a part. Yet what world would that be in Laszlo's view? With Ellul one can, I believe, affirm correctly that that cannot be the world of 'natural systems', as Laszlo wants to suggest; the subject must be relocated in the technological system. Certainly there is a shift of emphasis in Laszlo from the subject to the system; but one has to observe something remarkable: the subject comes face to face with himself again in the system, namely, as its constructor. Hence the kind of systems ethics advocated by Laszlo continues to be an ethics stamped by western subjectivism. In this normative ethics, the primacy of the autonomous subject is maintained, via the roundabout way of the system.

With Laszlo, however, one can raise equally well founded objections to Ellul's standpoint. Ellul defends the freedom of the subject and an 'ethics of freedom', but the norms for such freedom remain unclear. In Ellul's view these norms cannot be derived from the system, but he does not answer the question concerning what then is to guide the subject. For championing freedom is a perilous enterprise, as the technological society has taught us. How should we then live in our global world of technological systems? We have already noticed that in Laszlo's view the reorientation our culture needs can benefit from a rediscovery of ways of thought from classical antiquity and from the traditions of the Far East. Often the question is not even posed concerning positive impulses that our modern culture might derive from a reorientation to the Judaeo-Christian tradition that is one of its own spiritual roots. Yet why would that tradition not present some forgotten possibilities valuable for our technological society?

In what follows I want first to elucidate, in 1 through 3, a number of the fundamental notions of a Biblical view of reality. Then, in 4 through 7, I will indicate the nature of the normative systems ethics that can be elaborated by extending these notions.

1. Fundamental in the first place to a biblical view of reality is the notion that reality is relational in character. Nothing is self-sufficient, existing of itself. All things are connected. All things exist in comprehensive interconnection with one another and, to put

it even more strongly, all things refer to one another. This is true of the nature that environs us, of the rivers, seas and mountains, but no less of man himself with his history and culture. However, there is more. The most important feature of the relational character of reality remains to be expressed. Namely, all things refer not only to one another but transcend themselves to point to the origin, to God who created all things. Thus the poet of the old psalms says (*The Holy Bible*, Psalm 19:1): "The heavens are telling the glory of God; and the firmament proclaims his handiwork." The unutterable wealth and diversity of all that is created displays the wisdom of the Creator: to speak again with the psalmist (*The Holy Bible*, Psalm 104:24): "O Lord, how manifold are your works! In wisdom you have made them all; the earth is full of your creatures."

2. This fundamental insight that reality, including man himself, is relational in character represents a radical break with every form of anthropocentrism, which is, indeed, as Laszlo has correctly observed, a basic trait of our western culture. In biblical perspective it is not man that is the center of all that exists. The world was called into being by his creating word and in him all things find their existence and end. "For from him and through him and to him are all things. To him be the glory forever. Amen." (*The Holy Bible*, Romans 11:36) In philosophical language the insight into the relational character of reality can also be expressed by the term 'meaning'. Reality exists as meaning, but not as meaning through and for man (anthropocentrism), as that would yield no answer to the question concerning the meaning of man himself and of his history and culture. Man does not give to the world around him and to his own existence its ultimate meaning; rather, he must discover its given meaning. Dooyeweerd (1953,I:4) expressed this succinctly in his philosophy when he said: "*Meaning* is the *being* of all that has been *created* and the nature even of our selfhood."

3. Insight into the relational character of reality is set in even sharper relief by an endeavor to discover what determines the being of things as meaning. What determines that reality is meaning? The Creator wanted the relation to his creation and, reversely, the creation's relation to himself, to be determined by his law. All that is created is subject to and is bound by the law of God. By placing reality under his law, God causes it to be as meaning. Things are upheld by his creational law and creational word so that each thing

is as it is, created according to its own nature. The word of creation is what has called things forth and makes them be what they are.

4. For a normative systems ethics, the preceding observations mean, as I have also explained elsewhere (Strijbos, 1994), that it is assumed that different normative structures exist for societal-cultural reality and for the all the variegated phenomena found in it. For example, a marriage is different from a relation between business partners, engaging in scientific research is different from teaching, a state is essentially different from a family, conducting a medical practice should be distinguished from running a commercial enterprise, and so forth. Each of these phenomena has a normative structure of its own that determines its typical character. Normative structures are constants which, for the rest, do not preclude diversity but rather serve to make it possible. For normative structures are not blueprints but route signs; normative structures can be worked out in reality by human action in many different ways. Moreover, the normative order of things persists and is maintained even if a culture neglects it or deliberately strives to destroy it. No matter how deformed or suppressed they may be, normative structures always prevail and impose themselves in one way or another on human experience. However broken or dislocated the actual situation of a culture accordingly may be, the given normativity always provides points of connection for a fundamental and radical recovery.

5. Applied to the situation of our technological society, this means that we can support Ellul's critical stance regarding the dominion of technology. We must add however that the liberation of the subject in the technological society can only be realized when at the same time the normativity proper to the various sectors of our existence is again allowed to prevail. The systems which form the infrastructure for everyday life in every field must allow room for the original normativity of the various sectors of life. The systems themselves and the goals set for them do not form the ultimate horizon for a normative systems ethics. That would lead to an ethics of adaptation and control such that people would adjust to the system. The normative systems ethics advocated here is not a derivative of technical systems; rather, it precedes these systems. These systems must be so appointed or transformed that they create room for the unfolding of that for which the system is intended. Systems ethics in

our technology-ridden society can therefore be characterized as an 'ethics of disclosure'. The organizational-technological systems must be so reshaped that they once again offer protection and support for their threatened normative core.

6. An 'ethics of disclosure' requires in the first place, then, that we seek out the normative core of a particular sector of society. With respect to the health care system, for example, the question arises concerning the internal meaning and normative structure of the field of medicine. For the complex system of education, an answer must be sought to the question of the meaning determining framework of education in all its various forms. We must abstract from the complexity of the modern social system in order thus to penetrate to the archetypal phenomena that underlie the various fields of medicine, education, traffic, trade and economy, etc. What, for example, is the archetypal phenomenon, the basic given, from which the entire medical enterprise has developed and around which the entire complex system is built? Pellegrino (1981, 1988) speaks of 'the fact of illness' as the constituting phenomenon for medicine and its normativity. The situation of being sick entails in the internal nature of the case an ethical appeal to others for help. The field of medicine exists to respond to this appeal and provide disinterested assistance irrespective of persons. It occurs to me that the archetypal phenomenon of education can be identified as 'the fact of desiring to know'. The knowledge intended involves from the outset more than an instrumental approach to the world. The desire to know, with all the differentiation that entails, is before all else a desire for orientation and a determination of man's position in his world. Education is a matter in the first place of gaining a deeper understanding of why all things exists, of the meaning of man and world. This 'fact of desiring to know' has its own proper normative structure, which should be determinative for the student-teacher relation and for what in didactics is called the educational or learning process.

7. A following step in the elaboration of an 'ethics of disclosure' is to take a closer look at the system which, as we put it before, must be draped like a coat around the normative core of the system. In any case, a response must be forthcoming to the question, also posed by Ackoff (1994:57) in his reflections on systems ethics: "Who are the responsible actors within the system?" and "What responsibilities,

precisely, do the actors have?" Systems are sustained by many actors, each of whom must be able to participate in his own proper responsibility and also be able to be held accountable. The 'managing' of systems is a multi-actor process. Hence a key question of systems ethics is, "How are the responsibilities of the various actors related to each other?"

Chapter 13

NORMATIVE SOURCES OF SYSTEMS THINKING: AN INQUIRY INTO RELIGIOUS GROUND-MOTIVES OF SYSTEMS THINKING PARADIGMS

Darek M. Eriksson

"FIRST DISCOVER:
'Does God Exist?' is the most important question of systems thinking. "
C.W. Churchman, 1987:139

1.   INTRODUCTION

A central feature of modernism was the development of science that brought with it the so-called scientific method. The latter's role was to provide prescriptions for the production of scientific knowledge, which was characterised as objective, i.e. value and subject independent, and possible to reproduce by other scientists when so required. Due to its success the scientific method has been employed for virtually all types of human affairs, whether bridges, airplanes, banking operations, industrial organisations, hospitals, or community planning. Various customised dialects of the scientific method may be found today, under labels such as Operations Research, Management Science, Systems Analysis, Systems Engineering, and Systems Thinking. Over the years different specialised sub-areas have emerged, such as logistics, industrial engineering, computer science, software engineering, information systems science, quality technique. Here they are labelled all as 'systems thinking', including their various underlying paradigms.

Inspired by the philosopher, the late C. West Churchman, one of the founders and main critics of management science, operational research and systems thinking, this chapter investigates the normative foundations and religious sources of systems thinking. In a life long inquiry Churchman has provoked systems thinking with his intriguing questions and remarks, guided by a key question: "is it possible to secure improvement in the human condition by means of the human intellect?" (Churchman, 1982:19). Churchman was preoccupied with an heroic search for the boundaries of human reason and for the sources of the 'Ideal and Ultimate Good'. As he put it: "the problem of systems improvement is the problem of the 'ethics of the whole systems'" (Churchman, 1968a:4). Churchman's project aimed at the conception of a scientific method that explicitly incorporated ethical and other norms. As a means for reaching this end he pursued holism and pluralism, as articulated in *The Systems Approach* (Churchman, 1968b), where different approaches to planning and problem solving compete, such as the rationalist and humanist. In *The Systems Approach and Its Enemies* (Churchman, 1979), rationality is confronted with politics, aesthetics, ethics and theology. In *The Design of Inquiring Systems* (Churchman, 1971) alternative inquiring systems compete to secure the validity of propositions. Here - but also suggested in one of his very early texts, *The Theory of Experimental Inference* (Churchman, 1948) - Churchman shows that there is no way to secure complete validity of rationality, all propositions are founded on some assumptions that are subject to human conviction and faith, which brings us to the topic of this essay. Based on the assumption that religious convictions unconditionally govern all human activities, including intellectual and scientific, this study addresses the religious foundation of four systems thinking paradigms.[1]

Each paradigm comprises various theories, models, methodologies, methods and techniques, designed to support the solving of problems, of particular or general type. These theories and methodologies - formed by Western societies, taught at universities around the world and used by problem solvers (managers, engineers, policy makers, etc.) - may all be considered as intellectual tools for problem solving to design or shape our society and lives. But these tools have normative implications, because all design implies a notion of how things ought to be instead of their present state. A key question may then be posed: What kind of ethical, normative guidance do these tools provide, and why? In answering such a question, an understanding of its theoretical foundation is necessary, that is its

---

[1] The conception of a Paradigm is associated with Kuhn's (1970) work. The formation of 'system thinking paradigms', such as Hard Systems Thinking, Soft Systems Thinking, and Critical Systems Thinking, were all affected by the thoughts of C. West Churchman.

fundamental assumptions about reality, knowledge, man, and the source of norms. This understanding can be obtained through the identification of the religious ground-motives that govern the various paradigms.

Methodologically regarded, a model of religious ground-motives is employed as a conceptual tool for investigation of these paradigms. The results show that (i) the analysed paradigms are founded on different ground-motives, and that progress in their development may be clearly distinguished; (ii) these paradigms either: (a) rest on assumptions that have inherent contradictions, (b) do not address explicitly the normative content and consequences of their propositions for the design of human affairs, or (c) have normative foundations that lack social contract; and that (iii) the model used here for paradigmatic analysis can help uncover assumptions that other analysis of systems thinking paradigms does not achieve. The main message of this study is therefore to emphasize the need for explicit normative considerations in the management and design of human affairs, and to further pursue Churchman's project of conception of scientific method that incorporates human values into the design process. We first present the tool of this investigation: the model of religious ground-motives. Then we present four systems thinking paradigms and their respective ground-motive. Next, the model of religious ground-motives is compared with Jackson's (1982) use of Burrell-Morgan's (1979) model for paradigmatic analysis in the analysis of the systems thinking paradigms. This comparison shows the latter model's inability to uncover the normative limitations embedded in the four systems thinking paradigms. The chapter ends with a discussion and concluding remarks.

2. THE MODEL OF RELIGIOUS GROUND-MOTIVES [2]

In his original philosophical work, the Dutch philosopher Herman Dooyeweerd (1894-1977) focused his attention on the conditions of theoretic thought (Dooyeweerd, 1955). He has shown that all human

---

[2] Dooyeweerd's model has been used in this investigation because it is the most elaborated theory of religious ground-motives that is known to us, and also because one of the investigated systems thinking paradigms is based on Dooyeweerdian philosophy. However, similar investigations of systems thinking paradigms, employing other normative analysis frameworks, may be desirable, leading to a foundation for fruitful debate, when the results of the various analyses are juxtaposed.

thought is unconditionally grounded in religious convictions. This means that a 'religious ground-motive' prompts all philosophical activity, as well as all human life. 'Religious' is here a technical term, which Dooyeweerd defines as: "... the innate impulse of human selfhood to direct itself toward the true or toward a pretended Origin of all temporal diversity of meaning, which it finds focused concentrically in itself." (Dooyeweerd, 1955,I:57). A religious ground is a transcendent motive that takes hold of a person's heart, fills, motivates and dominates his every action - whether consciously or not. A religious ground-motive is said to be a "... moving power or spirit at the very roots of man, who so captured works it out with fear and trembling, and curiosity." (*ibid.*:58).

Dooyeweerd's histo-graphical studies of western philosophy laid bare the basic religious commitments of theoretic thought. He has identified four religious ground-motives, all of which identify the religious nature of philosophical activity in the western world. Three of the four ground-motives are 'apostate' - i.e. they have abandoned religious faith - and are said to contain an inner tension. All four religious ground-motives are presented briefly below, yet the investigation will show that only two of these will be pertinent for its result. All four are presented as they have contributed to each others' emergence, and presentation of all will provide increased understanding of the two employed in this analysis.

2.1   The Matter-Form Ground-Motive

The first of the four religious ground-motives is called matter-form. It is said that the dialectic of this ground-motive characterises all pagan Greek philosophy, and thus should not be associated exclusively with Aristotelian philosophy. During the early period of Greek history worship centred essentially on natural powers, thus the Greek religion was a nature-religion. It implied a worship of a formless stream of life out of which generations of beings periodically emerged. These beings were all subjected to death, fate and decay. This includes a continuous process of coming into being and passing away. The stream of life can only continue if individuals at the end of their allotted life are absorbed again. Individual man and beings are doomed to die and decay so that the cycle may continue. At a later stage a new type of religion arose. This was a culture-religion, as represented in the Homeric gods dwelling on Mount Olympus. These gods had left mother earth with her eternal cycle of life and death, and acquired a personal and immortal form of splendid beauty. They became gods of abiding form, measure and harmony.

Combined, these two religions - the nature-religion and the culture-religion - gave rise to the inner dialectic of the Greek matter-form motive. The nature-religion contributed to the principle of matter, that is, mortality and change, the elements of unpredictable mystery and the formless dark. The culture-religion, on the other hand, contributed to the principle of form, i.e. abiding being, light and heavenly splendour, as well as reason. According to Dooyeweerd (1955,I), these two principles controlled all Greek thought.

2.2   The Creation-Fall-Redemption Ground-Motive

The second religious ground-motive, probably the most familiar, is the Biblical one of creation-fall-redemption. This ground-motive dominates the Scripture and constitutes the Archimedean point that determines all Christian activity. The creation-fall-redemption relation is set by God as Creator, who gives His law to which creation is subordinated. In the Fall, humans separate themselves from God. Redemption is thus required to allow a full restoration and reintegration. Thus, this ground-motive is the religious presupposition of any theoretical thought that may claim a Biblical foundation. The Christian ground-motive is beyond the reach of a theoretical investigation, says Dooyeweerd.

2.3   The Nature-Grace Ground-Motive

The medieval synthesis that combined Biblical themes with Pagan thought is characterised by the third ground-motive, the nature-grace ground-motive. This ground-motive is a product of a synthesis mentality that adopted the tension-ridden matter-form ground-motive of the ancient Greeks and incorporated it into its nature motive. This implies that the notion of philosophy - including natural theology - belongs to the lower level of nature, while revealed theology belongs to the higher level of super-nature, or 'grace'. This articulation of dualism contains on the one hand a view of God as a transcendent pure form, and on the other hand a non-transcendent material world.

This synthesis philosophy culminated in the great Scholasticism of the Middle Ages. The nature-grace ground-motive came to its most articulate expression in the philosophy of Thomas Aquinas. It stated that human nature is weakened by the Fall and directed by a common natural law and the natural light of reason. Christianity, the Bible, faith, all these are specially added items. The natural man is not the radically fallen but a man endowed with reason that is one and the same in all men, and therefore the basis of all common neutral and autonomous areas of life. In the nature-grace ground-motive there was an implicit possibility of secularism because, if it is true that there is a whole realm of nature that possesses a certain amount of autonomy, then there is nothing to prevent the area of nature from going alone - there is no reason why whole areas of life may not be secular. Aquinas' position though, was that the realm of nature was the primer and necessary step toward the realm of grace. Hence, there was a necessary link and order between the two realms. William of Ockham was however determined to show that this link was not necessary. Ockham drove a wedge between these two areas in order to drive them as far apart as possible, which implied that the area of grace - i.e. of Bible and faith - was considered to have nothing whatsoever to do with the area of nature, i.e. the state, society, science.

2.4   The Nature-Freedom Ground-Motive[3]

While the philosophy of the Middle Ages was based on, and determined by the nature-grace ground-motive, it can be said that the modern period has adopted the Nature-Freedom Ground-Motive (Dooyeweerd 1955,I). As soon as the realm of grace was eliminated the realm of nature was accompanied by the realm of freedom, of autonomous man. However, this replacement of the realm of grace by the realm of freedom gave rise to a conflict between nature and freedom. This is so because nature is conceived as a set of scientifically discerned mechanical laws and processes which gives no room for autonomous freedom. Hence one of the problems of modern philosophy was to reconcile the two, that is, to answer the question: How can man maintain his autonomous freedom in a mechanically determined world?[4]

Immanuel Kant may be considered to be the father of the nature-freedom ground-motive. His Critiques attempted an idealistic solution to the nature-freedom tension. Kant's (1998) *Critique of Pure Reason* investigated the conditions determining the possibility of human knowledge. One of the central postulates is that the impressions humans receive from external objects are shaped and determined by an a priori structure of our mind. Kant rejected thus the long-standing assumption

---

[3] See also chapter 14, section 2.2 for more on the Nature-Freedom Ground-Motive.

[4] Dooyeweerd saw the history of philosophy as swinging back and forth between two antithetical poles within the various apostate ground-motives.

that human knowledge depends on external objects and replaced it with the claim that objects depend on our knowledge. However, the ultimate conditions or principles of knowing - that Kant called 'ideas' - are said to exceed the possibility of human experience. People could search for soul and world, but would never find them. God, the soul, immortality, freedom, as such cannot be known. Therefore, the best man may do is to act as if he had already reached those ultimate principles, and allow these regulative ideas to guide his further investigations. In his second major work - the *Critique of Practical Reason* (Kant, 1997) - Kant explored the conditions of moral actions. He stated that man's morality is essentially a matter of obligation and freedom. Kant's categorical imperative implies that individuals are capable of formulating laws of conduct for themselves and that the fundamental law is that one should perform an act only if one wished that anyone in similar circumstances would do the same thing. Moral law is a direct result of human reason, and moral goodness is determined by nothing other than reason. Right and wrong rest in man's autonomy, man is a rational being who knows a priori what is good without any help from outside. In order to reconcile the tension between the realm of nature and the realm of freedom, Kant wrote a third critique, the *Critique of Judgement* (Kant, 1987). There he attempts to construct a bridge between the two realms with the help of imagination.

3. FOUR PARADIGMS OF SYSTEMS THINKING AND THEIR RELIGIOUS GROUND-MOTIVES

System thinkers articulate often systems thinking in terms of various paradigms (Eriksson, 1998). The three most typical are Hard Systems Thinking, Soft Systems Thinking and Critical Systems Thinking. These three and the Multimodal Systems Thinking paradigm are investigated below in regard to their respective founding religious ground-motive.

3.1 Hard Systems Thinking: presentation and analysis

There are usually four distinct systems approaches associated with Hard Systems Thinking (HST) (Flood and Jackson, 1991b): operations research (e.g. Churchman, Ackoff and Arnoff, 1957) systems analysis (e.g. Atthill, 1975), systems engineering (Jenkins, 1969) and systems dynamics (e.g. Forrester, 1961). Even though diversity is present in particular theories, methodologies and methods related to these four approaches, they all have certain common properties that make it meaningful to organise them into one scientific paradigm. Ontoepistemologically they are all founded on some version of the realist-positivist position. They provide support for solving problems mostly by means of mathematical-statistical tools. They hold an assumption that an optimal solution exists and may be uncovered. Hence, the models are deterministic and/or stochastic as inherited from classical mechanics and thermodynamics. These systems approaches, stemming from the natural sciences, aspire to describe, explain and predict a studied phenomenon by means of the analytic-mathematical method. HST does not provide any explicit ethical-normative support for problem management although it seems that it is often employed together with the utilitarian approach - both aspire to optimise. HST is useful in situations where problems are already defined, and where their properties may be represented easily and meaningfully in numerical form.

This notion corresponds clearly to the nature realm of the Nature-Freedom Ground-Motive, as HST does not provide any account of human freedom or autonomy, intentions or non-rationalistic behaviour - such as emotional and spiritual - when referring to the axiomatic analytic-rational epistemologies. This implies that the realm of freedom is not only ignored and omitted, but also conditioned by the principles of the nature realm. Further, the methodologies of HST consider not only that it is possible to manage and control human affairs by means of the method of the realm of the nature but also that this is a preferred approach for such a purpose.

3.2   Soft Systems Thinking: presentation and analysis

Soft Systems Thinking (SST), (e.g. Jackson, 1982,1991) is often associated with methodologies such as Interactive Planning (Ackoff, 1981), Strategic Assumptions Surfacing and Testing (Mason and Mitroff, 1981), Interactive Management (Warfield and Cardenas, 1994), and Soft Systems Methodology (Checkland, 1978,1981).

One of the main distinctions between HST and SST is the difference in their respective onto-epistemological positions. SST tends to have a non-realist and non-positivist position, which varies between nominalism, interpretivism, constructivism, and pragmatism. This difference, which it seems to have inherited from the social and mind sciences, has important theoretical and methodological consequences. SST does not assume that problems exist and are well defined and independent of observers, and that they always have an optimal solution. Instead, problems are interpretations and/or constructions by its perceivers and/or conceivers, whose cultural backgrounds determine the understanding of a problematic situation. These situations are seen as 'messes' or networks of problems, which may be managed satisfactorily yet very seldom optimised. Diverse perceptions of situations should be accommodated with each other by means of communication and learning. Several of SST's approaches provide representational concepts peculiar to social systems, for example, owner, customer, and Weltanschauung. Hence, while HST's central issue is to explain and solve problems - as inherited from natural science - SST's central issue is to interpret and manage situations - as inherited from social science. SST arose from critique of HST approaches as delivered, for example, by Churchman (1970), Ackoff (1979a) and Checkland (1985), and took its present shape in the 1970's and 1980's.

SST recognises the autonomy or freedom of the human being, often conceived in teleological manner. Like HST, SST does not provide any explicit normative ethical framework as a guide for problem management. Instead, it relies on the value systems of the designers and stakeholders involved. Therefore it seems correct to consider SST to be founded on the realm of freedom, within the Nature-Freedom Ground-Motive. As a consequence of this lack of a normative framework, or perhaps as a reason for it, SST does not provide any solution to the tension between the realm of nature and the realm of freedom, other than considering SST - i.e. the realm of freedom - to be superior to and controlling HST - i.e. the realm of nature. HST is thus considered to be a special case of SST, which will decide where HST should be employed.

3.3   Critical Systems Thinking

3.3.1   Presentation of Critical Systems Thinking

Like SST, Critical Systems Thinking[5] (CST) emerged from critique of its forerunners, namely from HST and SST. While the critique of HST is similar to that formulated by SST, CST states that, in assigning itself primacy over HST approaches, SST is isolationist. Further, it is said that SST is unable to handle power structures in social situations, and that it is theoretically dogmatic and therefore not self-reflective (Flood and Jackson, 1991b). As a solution to these problems, CST has adopted Habermas' (1984,1987) version of Critical Social Theory. CST's programme has three main commitments, which reflect well the critique

---

[5] The author is aware that according to Ulrich (2003) a tension exists within CST between its two strands: Critical Systems Heuristics and Total Systems Intervention. The present analysis, however, investigates the very foundations of CST, and is valid for both strands.

that is delivered on SST: (i) theoretical and practical complementarism, (ii) promotion of emancipation, and (iii) critical self-reflection.

Theoretically considered, CST accounts for three kinds of rationality or inquiry. First is the empirical-analytical science that focuses on instrumental reason, provides nomological causal knowledge, and aims at explanation, prediction and control of nature, which is the sole foundation of HST. Second is the historic-hermeneutic science that provides a practical understanding of other human beings, on which SST is founded. Third are the critically oriented sciences such as psychoanalysis and Critical Social Theory that aim to provide emancipatory interest in freedom and overcoming unconscious compulsion. So far, CST offers the practitioner two methodologies; Critical Systems Heuristics (Ulrich, 1983, 1987), and Total Systems Intervention (Flood and Jackson, 1991a). The former aims to manage normative distribution in a design situation and therefore also manage social power structures in the design process. This is achieved by interrogating the conceptual boundaries or assumptions of a system design, through contrasting its 'is' with its 'ought', where all stakeholders are supposed to be represented in the design situation. It is Ulrich's work - inspired by Churchman - that has made an unparalleled breakthrough in systems thinking, by explicitly and operationally addressing the normative content of system design. Hence, unlike HST and SST, CST does support the systems designer with mechanisms for normative guidance. Even though CST first emerged in the 1970's, it did not become widely recognised until the late 1980's and the early 1990's.

3.3.2   Analysis of Critical Systems Thinking

CST is founded mainly on Kantian philosophy and Habermas' notion of Critical Social Theory. Compared with HST and SST, CST makes a serious attempt at representing the complete Nature-Freedom Ground-Motive. It not only articulates explicitly the two realms but also attempts to provide a link between the two. Thus, the empirical-analytic sciences represent the realm of nature, the historic-hermeneutic sciences represent the realm of freedom, and the critically oriented science represents the link between the two realms, aiming to secure a suitable co-existence. However, as CST stands in the Kantian shadow, it has not succeeded in solving the very fundamental tension between the realm of nature and that of freedom, which is: How can man maintain his autonomous freedom in a mechanically determined world? To make the argument more intelligible, the next few lines summarise Dooyeweerd's critique of Kant's position.

A starting point is that Dooyeweerd has shown that "All meaning is from, through, and to an origin, which cannot itself be related to a higher origin." (Dooyeweerd, 1955,I:9). Further, he has shown that truly critical thinking must accept the three transcendental Ideas, which in their trinity must be considered as the transcendental Ground-Idea (*ibid.*:89). These three are the Idea of the Universe (which Kant reduced to the sphere of nature), the Idea of the Ultimate Unity of Human Selfhood, and the Idea of the Absolute Origin. "Truly reflexive thought, therefore, is characterised by the critical self-reflection as to the transcendental Ground-Idea of philosophy, in which philosophic thought points beyond and above itself towards its own a priori conditions with and beyond cosmic time.", i.e. towards its origin (*ibid.*:87). Further, Dooyeweerd has shown that "A sharp distinction between theoretical judgements and the supra-theoretical pre-judgements, which alone make the former possible, is a primary requisite of critical thought." (*ibid.*:70). Consequently, "Every philosophical thinker must be willing to account critically for the meaning of his formulation of questions. He who really does so, necessarily encounters the transcendental Ground-Idea of meaning and of its origin." (*ibid.*:70). The Kantian system of thought - and therefore Habermasian notion of Critical Social Theory and thus Critical Systems Thinking - never fulfilled the latter requirement. Instead, their thinking is founded upon the yet unresolved dogma of the Nature-Freedom tension. More specifically, the dogma of Kantian epistemology is that it issues from the assumption of the autonomy of theoretical thought (*ibid.*:35). Kant did not accept the three transcendental Ground-Ideas as the real hypothesis of critical philosophy. Therefore, he could not see that "... in their very theoretical use, they must have a real content which necessarily depends upon supra-theoretic pre-suppositions differing in accordance with the religious ground-motives of theoretic thought." (*ibid.*:89). Instead, "He [Kant] restricted their significance theoretically to a purely formal-logical one; they have, according to him, only a regulative, systematic function in respect to the use of the logical concepts (categories) which are related apriori to sensory experience." (*ibid.*:89). When asking why Kant, at this critical point, abandoned the real transcendental motive, Dooyeweerd states that Kant "... had become aware of the unbridgeable antithesis in the ground-motive of nature and freedom, and now rejected every attempt at dialectical synthesis." (*ibid.*:89). More precisely, the problem is that Kant did not see "... that his theoretical epistemology itself remained bound to a transcendental Ground-Idea, whose contents were determined by this very basic motive." (*ibid.*:89).

In the line of this argument, Strijbos' (1995) investigation of various systems approaches concluded that CST does broaden the notion of rationality when compared with the earlier approaches, which implies that the normative and ethical question is assigned a central position within CST. However, Strijbos (1995:373) emphasises that rationality as such "... retains its place as the necessary, all-embracing framework for approaching reality." The problem is thus, that: "Science as an instrument of control is subjected to criticism by 'critical systems thinking' but this critical thinking remains subject itself to an autonomous rationality." (*ibid.*:373-374).

3.4   Multimodal Systems Thinking: presentation and analysis

Multimodal Systems Thinking (MST), (e.g. De Raadt, 1991,1997a) is the most recent contribution to the family of systems thinking paradigms here presented, and probably not yet generally accepted. MST has its foundation in the Cosmonomic philosophy of Dooyeweerd (1955) which is founded in the Christian Reformation - the latter provides it with the source of its normative content and guidance.[6] A second central theoretical component of MST is Multimodal Theory which has been inherited from Dooyeweerd's (1955,II) General Theory of Modal Spheres. In this respect, Dooyeweerd maintained that human thought is based upon and bound to its experience, and that this experience exhibits a number of modalities, aspects, or spheres of law. MST combines Multimodal Theory with, among others, General Systems Theory and management cybernetics. Methodologically considered, this approach offers the Living Social System Model (De Raadt, 1991) and Multimodal Methodology (1995); both aiming to support systems design in a broad manner. In De Raadt's (1996) notion, MST "... is based upon the presupposition that the universe is ordered and that this order encompasses the totality of natural phenomena and human life. This implies that there exists a truth that is absolute and autonomous from man and nature."

The criticism levelled by MST is similar to the critique delivered by SST. When it comes to SST and CST, both are considered to be normatively unsatisfactory. SST is said to promote nihilism while CST is

---

[6] More recently another dialect of systems thinking founded on Dooyeweerdian philosophy emerged, 'Disclosive Systems Thinking', which is presented in chapter 14.

said to be founded on a dogma that assumes human reason to be autonomous. According to MST such a standpoint is unable to provide an appropriate normative guidance for human affairs. This is the main argument against CST, where MST distinguishes itself by providing an external source of norms: the Christian Faith, which is assumed to transcend human autonomy.

MST rejects the intellectual supremacy of rationality, where the latter is conceived to be embedded within a transcendental normative order that determines the status of reason and science. Reason is considered to be dependent upon and determined by faith. God has provided the universe with laws that govern it. Their relation to each other is expressed in the Multimodal Theory, which is meant to be a theoretical abstraction of the human universe that is governed by both determinative and normative laws. MST thus claims to be founded on the religious ground-motive of creation-fall-redemption.[7]

An issue of the Biblical foundation is to judge which interpretation is to be considered valid. The Christian community itself has not succeeded in finding a common unity. In contemporary Western societies an unconditional faith in God and Jesus Christ is not widely accepted. Table 1 shows an overview of the investigation carried out in this section.

Table 1. Shows the religious ground-motives of the four investigated systems thinking paradigms. Hard Systems Thinking was found to articulate the realm of Nature of the Nature-Freedom Ground-Motive and Soft Systems Thinking, the realm of Freedom. Critical Systems Thinking was found to articulate the whole Nature-Freedom Ground-Motive, while Multimodal Systems Thinking was found to articulate the Biblical motive of Creation-Fall-Redemption.

	Paradigms of Systems Thinking			
	Hard Systems Thinking	Soft Systems Thinking	Critical Systems Thinking	Multimodal Systems Thinking
Dominating Religious Ground-Motive	Nature	Freedom	Nature - Freedom	Creation - Fall - Redemption

---

[7] A critical thesis put forward by Strijbos in a personal communication is that MST is not purely founded on the ground-motive of creation-fall-redemption because it has merged Dooyeweerdian modal theory with theoretical components from Stafford Beer, which are founded on the nature-freedom ground-motive. Hence MST may have a fundamental flaw. This thesis is beyond the scope of this chapter. [But see chapter 14. Eds.]

4.   A COMPARISON OF THE MODEL OF
       RELIGIOUS GROUND-MOTIVES WITH
       JACKSON'S EMPLOYMENT OF THE
       BURRELL-MORGAN MODEL

Given the results of this paradigmatic analysis, a reasonable question is: What is its value? In order to provide one point of reference for such an assessment, a comparison with another paradigmatic analysis is made.

Several models have been advanced with the purpose of investigating the variety of approaches of systems thinking (e.g. Ulrich 1983, Miser and Quade 1985, Bowen 1986, Eden and Radford 1990, Jackson 1991, Strijbos 1995, Flood and Romm, 1996b). One popular approach is Jackson's (1982, 1991) use of the Burrell and Morgan (1979) model of sociological paradigms to interpret the various contributions to systems thinking. This section presents a comparison between Jackson's investigation and the present one, to identify differences between the two models, and thus illuminate the respective strengths and weaknesses of the model of religious ground-motives and suggest the value of the paradigmatic analysis presented in this essay. A summary of Jackson's analysis is presented prior to the comparison.

Burrell and Morgan's (1979) model of sociological paradigms attempts to illuminate the assumptions on which the various theoretical models within social science are founded. The model is a matrix along two variables. One is about the nature of social science, and the other is about the nature of social systems. The first variable may take an objectivist or a subjectivist position. An objectivist position implies realist ontology, positivist epistemology, determinist anthropology and nomothetic methodology. A subjectivist position implies, on the other hand, a nominalist ontology, anti-positivist epistemology, voluntarist anthropology, and ideographic methodology. The second variable implies assumptions about the nature of society. One position is the sociology of regulation of status quo social systems, while the other is the sociology of radical change, which considers social systems to be conflicting and dynamic. Burrell and Morgan combined these positions and thus obtained four possible sociological paradigms. These are the Functionalist paradigm (objectivist-status quo), the Interpretive paradigm (subjectivist-status quo), the Radical Humanist paradigm (subjectivist-dynamic), and the Radical Structuralist paradigm (objectivist-dynamic).

The Functionalist paradigm implies that the studied systems are easy to identify and describe, and possess an existence independent of its observers. Their study searches for regularities and relationships between the various components. Human behaviour is determined by its environment, and the studied system as such is characterised by status quo. Quantitative models are built as representations meant to facilitate prediction and control of the studied systems.

Second, the Interpretive paradigm implies that there are individual interpretations of the observers, which may well vary according to the observer. The voluntarism of humans makes it in practice very hard to construct a feasible quantitative model; the constructed models are qualitative where knowledge is obtained by involved studies. Still the idea of the study as such is to identify some status quo, so that understanding and eventual prediction and control may be obtained.

Thirdly, the Radical Structuralist paradigm postulates an independent existence of the studied social reality. It searches for regularities of deterministic system behaviour. This implies a representation of radical changes and conflicts.

Finally, the Radical Humanist paradigm considers the social reality to be a construction of its observer, which implies that personal involvement is necessary to question these systems, and qualitative representation will be its result. Again there is a notion of transformation and change of the social systems.

Jackson (1982,1991) allocated the various systems thinking approaches into this model of sociological paradigms, as follows. The approaches labelled Hard Systems Thinking, in other words Operations Research, Systems Analysis, Systems Engineering, and Systems Dynamics, were found to belong to the Functionalist paradigm. Secondly, approaches belonging to Soft Systems Thinking, that is Soft Systems Methodology, Interactive Planning, and Social Systems Design, were allocated to the Interpretive paradigm. Thirdly, Critical Systems Heuristics of Critical Systems Thinking were found to be Radical Humanist in orientation. Radical Structuralism was found to have no relating systems thinking paradigm. See Table 2 for an overview.

What is the value of the model of religious ground-motives for paradigmatic analysis of systems thinking paradigms? The religious ground-motive model shows a limitation of the Burrell-Morgan model itself and therefore of any positioning of systems thinking paradigms within that model. The Burrell-Morgan model itself is grounded on the Nature-Freedom Ground-Motive, which is manifested in the assumptions of dualism in the nature of Social Science. This both articulates and forces the investigated systems thinking paradigms into the unbridgeable tension of dualism, founded on the assumption of autonomous reason. Realist ontology, positivist epistemology, determinist anthropology and nomothetic methodology express perfectly the Nature Realm while nominalist ontology, anti-positivist epistemology, voluntarism anthropology and ideographic methodology express the Realm of Freedom. A problem with the use of the Burrell-Morgan model for paradigmatic analysis is thus that, since it itself is founded on the Nature-Freedom Ground-Motive, it does not allow the detection of problems in the very ground-motive that governs many of the systems thinking paradigms. A second problem is that this model mis-conceptualises systems thinking paradigms that are not based on the Nature Freedom Ground-Motive, such as Multimodal Systems Thinking. Finally, Jackson's analysis does not inquire explicitly the sources of norms of these paradigms.

Table 2. Illustrates Burrell and Morgan's (1979) model of sociological paradigms together with some systems thinking approaches, as allocated by Jackson's (1982, 1991) investigation. Explanation of acronyms: OR: Operations Research, SA: Systems Analysis, SE: Systems Engineering, SD: Systems Dynamics, SSM: Soft Systems Methodology, IP: Interactive Planning, SSD: Social Systems Design, CSH: Critical Systems Heuristics.

Radical Change Sociology

	Radical Humanism CSH	Radical Structuralism - empty -	
Subjective social science	Interpretative SSM, IP, SSD	Functionalism OR, SA, SE, SD	Objective social science

Regulation Sociology

5.   SUMMARY AND CONCLUSIONS

This investigation starts with the conviction that it is important to understand the intellectual and pre-intellectual (i.e. religious) foundations of systems thinking paradigms. The reason for this is that each paradigm consists of theories, models and methods employed for problem solving in human affairs. This implies that the use of these theoretical constructions in problem solving influences the content of the generated problem solutions and hence the design of how human affairs ought to be. The latter implies necessarily that a normative decision is made, when making a choice of how something ought to be. Given that all norms are ultimately based upon religion - whether on Christianity, Judaism, Islam,

Buddhism, or any other - the question of this investigation was: which religious ground-motives govern systems thinking paradigms? The results show that:

♦ The Hard Systems Thinking paradigm is based on the Nature Realm of the Nature-Freedom Ground-Motive;

♦ The Soft Systems Thinking paradigm is based on the Freedom Realm of the Nature-Freedom Ground-Motive;

♦ The Critical Systems Thinking paradigm is based on the whole Nature-Freedom Ground-Motive;

♦ The Multimodal Systems Thinking paradigm is based on the Creation-Fall-Redemption Ground-Motive;

♦ A progress or evolution in time of systems thinking paradigms may be identified; starting with HST, hence the Nature Realm, through SST, hence the Freedom Realm, and currently ending with CST, hence the Nature-Freedom Realms, and MST the Creation-Fall-Redemption motive.

This means that HST, SST and CST are founded on religious convictions containing an inherent contradiction, implying that these paradigms do not offer a coherent and stable foundation for normative guidance. MST, on the other hand, is founded on a coherent and stable religious conviction, which does offer normative guidance. However, this guidance seems to lack the social contract necessary for its employment in practice. Consequently, this means that none of the systems thinking paradigms, here investigated, provide viable normative guidance. This can open various questions, such as: What should characterise such an approach? Is it possible to formulate such an approach? Or even: Is it desirable to do so? Finally, in order to assess some aspects of this investigation's value, the employed model of religious ground-motives was compared to another model of paradigmatic analysis. This is Jackson's use of the Burrell and Morgan model of sociological paradigms. The result of this comparison shows that: (a) the very model of Burrell and Morgan is founded on the Nature-Freedom Ground-Motive, hence on an theoretical antinomy; (b) as a consequence of this, its analysis of systems thinking paradigms mis-conceptualises the foundations of these paradigms with regard to their normative foundations, when compared with the model of religious ground-motives. The value of the present investigation resides in the fact that new knowledge about systems thinking paradigms has been obtained; knowledge that is, in our opinion, of importance because it searches for the ability to set the direction for our managerial decisions and actions.

Chapter 14

TOWARDS A
'DISCLOSIVE SYSTEMS THINKING'

Sytse Strijbos

1. INTRODUCTION

In Chapter 6 it has been pointed out that one of the principal distinguishing features of modern society is modern technology. By that I do not mean that in their everyday world people are surrounded by countless material artifacts, for that was the case even in pre-modern societies. Nor may the epithet 'technological society' be taken simply to indicate tools, machines and technological things as separated objects that confront human beings in their habitat. The reference is rather to the habitat itself. The multiplicity of technological things and processes that surround us form together an entirely new environment in which human life now unfolds in the world. Technology as a distinguishing feature of modern society means that in their everyday activities people have become dependent on numerous technological-organizational systems for communication and transportation, food production and distribution, health care and much more.

In Chapter 12 the idea of a systems ethics as an ethics for our technological society has been discussed. This society confronts us in many areas with questions of design and management of the technological-organizational systems or infrastructures within which modern life unfolds. At issue are questions like: What societal agents are responsible for particular developments? How are the different responsibilities of the agents related to each other and how are they coordinated? What are the norms for action by the various agents?

The present chapter turns attention to the approach or methodology, which is applied in the mentioned processes of structuring or restructuring of our technological world. The following section starts with an evaluation of current systems methodologies focussing on the relation between systems ethics and systems methodology. Against that background, a proposal is made in section 3 for a 'disclosive systems thinking'[1] leading to four normative principles guiding its practical application. Finally, section 4 offers some concluding remarks pointing to further research in 'disclosive systems thinking.'

2.   AN EVALUATION OF CURRENT SYSTEMS METHODOLOGIES

With respect to the question of the connection between systems methodology and systems ethics, one might all too easily be led to say that it is external. Underlying this answer is the notion that the field of systems methodology is located next to but apart from the field of systems ethics. These two fields are thus entirely separate and distinct, and systems methodology, thus regarded, is viewed as an (ethically) neutral matter. In the continuation I want to advocate a different perspective, namely, that any systems methodology implies, whether conspicuously or not, a particular normative idea of systems ethics. According to this view, systems ethics is thus not just added on afterwards but is implicit from the outset in whatever systems methodology one may adopt. To make this clear, I want to begin in 2.1 by saying something briefly about systems methodology. What are we to understand by that? Next, I will investigate in 2.2 the relation between systems ethics and the different strands of systems thinking. To what kind of systems ethics are these strands oriented, and how is that justified?

2.1   Systems methodology entails systems ethics

In the first pages of his book *Systems Methodology for the Management Sciences* (1991) Jackson presents an explanation of the term 'systems

---

[1] What is here called 'disclosive systems thinking' has a family relationship with 'multimodal systems thinking', discussed in other chapters of this volume. Both have borrowed fundamental ideas of the philosophical work of Dooyeweerd. However, there is a crucial difference. Multimodal systems thinking merges Dooyeweerd's theory of modalities and Stafford Beer's cybernetic theory of management. Disclosive systems thinking follows a more radical strategy by focusing on the underlying ontology and philosophical underpinnings of systems methodology. It is rooted in Dooyeweerd's notion of disclosure.

methodology'. It can be broken down, so he correctly observes, into two questions: What kind of methodology is at issue? and What is meant by the adjective 'systems'? Jackson states that the word 'methodology' can be used in two senses: (1) in the social sciences it pertains to the scientific method as the handmaiden of theory formation, which is to say to the procedures and rules that one must apply in order to gain theoretical knowledge of social reality; however, (2) 'methodology' can also refer to practice, or to the solution of the problems that arise in social reality.

It is worth noting that Jackson relativizes again in the continuation of his argument the clear and valuable distinction between the two senses of 'methodology' with which he begins. Instead of a difference in principle there is as Jackson (1991:3) sees it only a gradual difference: "The two meanings given to methodology in the social sciences and systems thinking suggest a different emphasis rather than a real distinction," he states. Given the two senses of methodology distinguished above, one can say that in Jackson (1991) the accent falls on the latter. He concentrates on systems intervention methodologies, because his main concern is with the practical task of managing problems and bringing about change.

So systems methodology in the service of the management sciences is practice-oriented since it is aimed at altering reality and concerns "the organized set of methods an analyst employs to intervene in and change real-world problem situations." Yet in intervening in the world around us we proceed in our methodology from certain theoretical assumptions concerning this world. Therefore we have according to Jackson (1991:4) "little option but to turn to ideas developed in the social sciences for the purpose of probing the theoretical underpinnings of methodologies." Systems methodology that is aimed at the solving of practical problems is thus also dependent, according to Jackson, as methodology, upon theoretical insights from the social sciences. The connection that is posited here between systems methodology, on the one hand, and social theory and philosophy, on the other, is indeed of crucial importance, and we shall return to it again in what follows. But this quite correctly signalized connection does not abolish the difference in principle, as Jackson sometimes seems to give the impression it may, between a methodology that serves theory formation and a methodology applied for purposes of intervening in concrete reality.

Methodology has as its object of study the scientific method itself and may therefore be defined as theoretical reflection on the scientific method. And systems methodology is thus concerned with the scientific method for approaching practical problems, where the notion of 'system' expresses that the approach is intended to be integral. One might even speak here of a scientific method of designing. Normally, designing suggests to us the activities of engineers. But that is too restricted. Thus Simon correctly pointed out that we encounter designing not only in technology but that it in fact forms the core of the various professions. In his *The Sciences of the Artificial* (1996, third edition:111) we read:

> "Engineers are not the only professional designers. Everyone designs who devises courses of action aimed at changing existing situations into preferred ones."

And Simon continues:

> "The intellectual activity that produces material artifacts is no different fundamentally from the one that prescribes remedies for a sick patient or the one that devises a new sales plan for the company or a social welfare policy for a state. Design, so construed, is the core of all professional training; it is the principal mark that distinguishes the professions from the sciences. Schools of engineering, as well as schools of architecture, business, education, law, and medicine, are all centrally concerned with the process of design."

It is not my intention to go into Simon and the issue of design broadly here. A few comments must suffice. Thus I agree with Simon that designing is a core activity in various professions. I also agree that what he calls the sciences of the artificial - think of the technological sciences - need to be distinguished from the natural sciences. The natural sciences are aimed at knowledge of the physical world, while the technological sciences embrace all scientific knowledge connected with the designing and realizing of technological objects and, more broadly, the structuring of our technological world. Yet while Simon correctly asserts the distinctive character of the sciences of the artificial as over against the natural sciences, he loses sight of some essential features of these sciences considered as normative, or cultural, sciences.

In the first place, it must be kept in mind that designing occurs on a theoretical basis but is not focused on reality as it is. Designing focuses on expressing insight into something new, on reality as it can become, or be made to be. Directly connected with this, in the second place, is the fact that norms are involved in the designing of something new. Designing is not a predictable or determinable activity, and it does not lead to insight into reality as it must be of necessity. In respect of the first point - that designing is not a determinable activity - the sciences of the artificial resemble the natural sciences. The investigations of a researcher in the natural sciences also transcend the scientific method and existing scientific knowledge in a particular field. How an ambitious, highly motivated doctoral candidate is to arrive at his or her hypotheses for research is not something that can be taught. And it is precisely that aspect of a candidate's project that often turns out to be the most difficult part of all.

We return to the discussion of systems methodology. We have seen that it is to be regarded as design methodology and that, as such, it thus involves theoretical knowledge of the scientific method of design. As is the case in every discussion of the foundations of a particular field, fundamental philosophical questions are at issue here. To gain a clearer view of these questions it will be helpful to draw a comparison with the scientific method of design in technology.

The parallels between technique, conceived in the narrower sense of the engineer's work and thus as the shaping of inorganic material with the help of tools (cf. Van Riessen, 1979a,b), and management can be broadly sketched from a systematic standpoint as in Fig. 1, which features four levels.[2] For technique, in the left column, these are, from top to bottom: technique as the practical activity of forming or shaping inorganic materials with the help of tools, followed by the level of the technological sciences and, below that, the level, again, of mathematics and the natural sciences as the foundation sciences for technology, followed by the level of philosophy and the corresponding world- and life-view. The column to the right is for management: management as a practical activity of organizing people and finding means for the realization of particular ends, supported by the level of management sciences and systems methodology, which in their turn are rooted in mathematics and the social sciences. Given the subject of this section, the relation between systems methodology and systems ethics, I want to make some notes about the relations between the levels (b), (c) and (d) of Fig. 1.

In general, an engineer is meant to have knowledge of physics and mechanics and of the properties of the materials to be shaped. While technology, in the case of designing a chemical reactor for example, is based on physical and chemical technological sciences, these technological sciences are in turn dependent on the foundation sciences of mathematics, physics and chemistry. Dependency implies here that in the domain of technological sciences, level (b), a 'translation' is needed, as it were, of the insights of the natural sciences and general philosophy and world- and

---

[2] What is referred to here as technique and management are respectively the narrow and broad concept of technology distinguished in Chapter 1, Fig. 1.

life-view, levels (c) and (d).[3] For the management sciences and systems methodology the situation is analogous. The task of the manager confronted with problem situations requires insight into the 'social material' impacted by management activity. Just as natural scientific knowledge must be translated into the language of technological science and the design methodology of the engineer, so in a similar manner social theory, layer (c), and the normative view on man and society entailed in the underlying general philosophy and view of the world, layer (d), must be converted into usable concepts and approaches for dealing scientifically with management problems, layer (b).

(a)	technique (in the narrower sense of the engineer's work)	management and organizational technique
(b)	technological sciences design methodology	management and organizational sciences systems methodology
(c)	mathematics natural sciences	mathematics social sciences
(d)	philosophy and world- and life-view	philosophy and world- and life-view

Figure 1. Parallels between technique and management

It is just here, at this point, that we find what we have been seeking: the inner point of contact between systems methodology and systems ethics. Systems methodology is connected with 'social material' and the normative structures proper to it. These inherent normative structures have implications for the development of scientific method and its application to concrete problem situations.

2.2   From 'Hard Systems Thinking' to 'Critical Systems Thinking' and beyond

In the second part of his previously mentioned book *Systems Methodology for the Management Sciences*, Jackson presents a broad survey of five systems methodologies. He deals with the organizations-as-systems tradition, several variants of 'hard systems thinking',

---

[3] For a broader orientation concerning the relationship between technology, technological sciences and the foundational sciences, see for example Rapp (1974), Mitcham (1994a).

'organizational cybernetics' with special attention for the work of Stafford Beer, 'soft systems thinking', and finally 'critical systems thinking' as a relative newcomer in the tradition of systems thinking. As noted earlier, Jackson uses the term 'methodology' in two senses. That is evident again in these designations. Thus Jackson (1991:68,69) observes in connection with the 'organizations-as-systems' approach, for example, that although this was developed primarily "as a methodology for understanding organizations", it is also applicable "as a methodology not for producing knowledge about organizations but for recommending action."

We will perhaps do well to notice here again, if only obliquely, that the distinction and the relationship between the two types of methodology are not made clear. It is true, to be sure, that one may often derive concrete courses of action from a theory, in this case a systems theory of human organizations. Nor can there be any objection to identifying as a methodology the procedures that are required to distil such plans of action from a theory. Such a methodology forms a bridge between theory and practice. My objection is to any blurring of the distinctive character of the scientific method as it is applied to the resolution of practical problems, on the one hand, and methodology as theoretical with respect to all that, on the other hand. Scientific method is used here not to construct theory or to build bridges between theory and practice but to prepare human interventions in reality. 'Hard', 'soft' and 'critical' systems thinking all refer as methodologies to the last category, the scientific method used to intervene or transform 'social material'.

In this article it is more important to consider how the development from 'hard systems thinking' via 'soft systems thinking' to 'critical systems thinking' is to be assessed. Or, more precisely, what conceptions of systems ethics and thus what normative principles are implicit in these systems methodologies? The perspective from which we want to respond to this question is given with the ideal of systems thinking, which has been the issue from the outset. Since the days of the founders in the fifties and sixties, a major current in systems thinking has endeavored to find an answer to the prevailing mechanistic-technical paradigm in science. As a counter to mechanistic-reductionist thinking in terms of closed systems, Von Bertalanffy launched the concept of the open system as a basis for a holistic science. As a pioneer in operations research Ackoff defended a systems approach as a counter to what he called 'Machine Age' thinking. Together with Churchman he gained the leadership, shortly after the Second World War, of one of the first systems groups in the United States, established in the Philosophy Department of the University of

Pennsylvania. With their systems approach Ackoff and Churchman strive to uphold the original intention of operations research as a holistic, interdisciplinary scientific method.

Comparable with these developments in North America is the work of Checkland and Vickers in England at the University of Lancaster (cf. Jackson, 1982, 1991). Instead of a 'hard' engineer's approach to practical problems "based upon the assumption that the problem task ... is to select an efficient means of achieving a known and defined end," Checkland (1978:109) seeks a scientific approach to "unstructured or 'soft' problems." "Here, one aspect of the problem situation is precisely the impossibility of defining desirable ends or objectives. The 'soft' systems thinking which is applicable in such situations has to abandon the means-end model upon which 'hard' systems thinking is based."

It is remarkable that from Von Bertalanffy's day in the 1950s to today the original dream of systems thinking has remained alive. In changing situations, the constant intention has been to break the dominance of the technical world picture in science and culture (Strijbos, 1988). New impulses were even provided by the work of one of the keenest critics of systems thinking, the German social theorist and philosopher Habermas. His critique of the dominance of technological categories in the process of development of modern society inspired a younger generation in England in the eighties to work out a program called 'critical systems thinking', a systems methodology based on Habermas's social theory. Although 'soft systems thinking' attacked the technological rationality embodied in 'hard systems thinking', one crucial element, so its critics observed, was never targeted. 'Soft systems thinking' proceeds from existing power relationships and "therefore merely facilitates a social process in which the essential elements of the status quo are reproduced." "Checkland, like Churchman and Ackoff," so their critics charge, "fails to give attention to coercive problem contexts." (Jackson, 1991:169,180)

The core of the criticism levelled by 'critical systems thinking' against existing soft systems methodologies is thus that in approaching problem situations these offer no perspective for the emancipation of people or for their liberation from repressive social structures. Such methodologies find their point of departure in the given social arrangements and accept the limitations these impose. Given their regulative orientation, they lack the potential to bring about radical change. In an extensive analysis and assessment, Jackson (1991:180) finally identifies soft systems thinking as "a rather underdeveloped form of critical modernism, based upon Kant's program of enlightenment and seeking the progressive liberation of humanity from constraints." And because it is so underdeveloped, "soft systems thinking is prone to slipping back into becoming no more than an adjunct of systemic modernism, readjusting the ideological status quo by engineering human hopes and aspirations in a manner that responds to the system's needs and so ensures its smoother functioning."

Before proceeding to explicate the normative systems ethics inherent in the successive systems methodologies, it is insightful to place them against the background of the tradition of Western thought. As has been shown by Dooyeweerd (1968, 1979) in his extensive analysis of Western thought there is an inherent tension in modern thinking since the Italian Renaissance of the 15th century. This is a result of the clash of two ideals, namely, the ideal of personality and the science ideal, which since Kant have in general been designated as the opposed motives between *nature* and *freedom*. Both arise from the same spiritual root. The humanist ideal of personality demands the complete autonomy of the human person, expressed in an ideal of freedom. However, the absolutization of human freedom calls forth a second ideal, the ideal of science. How else can human freedom be realized than through a liberation of the person from all ties and through active control of the world around him? And control is possible thanks to the scientific understanding of reality.

The personality ideal summons up the science ideal. Yet when one lives consistently by both, they seem inevitably to become opposites. Human freedom is at risk of being destroyed rather than confirmed by human scientific intervention in reality aiming to set people free. This tension between the two poles of freedom and control manifests itself through the whole history of modern Western thought, including as I would point out here the recent history of the systems sciences since the fifties of this century. One might say that 'hard systems thinking' is oriented to the pole of control, and the notion of a systems ethics that underlies it might be called an 'ethics of adaptation and control'. 'Soft systems thinking' tries to make a shift to the opposite pole of freedom but still has a regulative orientation. In contrast, 'critical systems thinking', which is also oriented to the pole of freedom, seeks radical changes at the systems level and is based on an 'ethics of liberation'. Established societal power relationships are not legitimated but criticized by the methodology.

It may not be denied that 'critical systems thinking' which shares Habermas' protest against what he called the monological Cartesian subject, tries to break, in a certain way, with the tradition of Western thought. In his debate with Luhmann, Habermas' criticism is that systems theory is an attempt to borrow and improve upon the basic concepts and problems of this out-dated philosophy of the knowing subject. The relation between the knowing subject and the object turns up in systems theory in another form, as the relation between a system and its environment (see Strijbos, 1988, ch. 3 and Bausch 1997). Yet notwithstanding the resistance Habermas registered against the dominance of cognitive-instrumental rationality in the Western world, he persisted in adhering to the idea of autonomy and an autonomous rationality. While directing the solitary subject to return to the dialogical world, Habermas chose his point of departure in an expanded notion of communicative rationality.

It is precisely regarding this point of the status of human rationality in Western thought that we seek to go another way. Decisive for it, as Dooyeweerd showed in his main work, *A New Critique of Theoretical Thought* (1969), is the relativizing of the autonomy of human rationality and the restoration of the subject to its original interconnectedness with reality. The whole of human existence, including human rationality, is tied to this and manifests itself as a response to the creaturely order given for it. For systems ethics this means that over against the poles of an 'ethics of liberation' and an 'ethics of adaptation and control' an alternative can be indicated that may be called an 'ethics of response and disclosure'.[4] In the latter view neither the subject nor the system is regarded as the final normative horizon for human action. 'Disclosive systems thinking' and the systems ethics entailed in it proceed from the normative view that the various systems receive their meaning from the pre-given reality and order of which these systems are a part. In other words, the idea of an intrinsic normativity is accepted as a leading principle for human intervention in reality and the endeavor to shape the world. Or, better: human action forms a response to this intrinsic normativity and may as such disclose structural possibilities that are enriching for human life and culture.

3.   NORMATIVE PRINCIPLES OF 'DISCLOSIVE SYSTEMS THINKING'

In the foregoing it became clear that a fundamental idea has remained unquestioned in these sciences to this day, namely that of human

---

[4] What is referred to here as an 'ethics of disclosure' is called elsewhere (Strijbos, 1996, Part II) an 'ethics of transformation'. At that time I could not find a better term although I realized its shortcomings because the term 'transformation' may cover different normative perspectives and does not make clear the difference from the other positions.

autonomy and especially the autonomy of human rationality. As said, it is just at this crucial juncture that 'disclosive systems thinking' seeks to go another way in the systems field: accepting the idea of heteronomy, 'disclosive systems thinking' aims to renew the underlying ontology and foundations of systems methodology. In 2.1. it has been explained how each systems methodology (cf. level b in Fig. 1) is based on two underlying levels. Fundamental ideas of a certain world- and life-view and related philosophical concepts (level d) are expressed at higher levels. For 'disclosive systems thinking' this leads to the formulation of four major commitments (section 3.1-3.4) or normative principles which may guide human action towards the unfolding of human life and culture in the context of our modern technological societies.[5]

Section 3.1 explains that creative human cultural activity evinces the character of a response. Primary for the development of human society and culture is the norm for the opening or *disclosure of everything in accordance with its inner nature or its intrinsic normativity.* Characterizing cultural formative activity as 'disclosure by response' leads to the identification of a second normative principle (discussed in section 3.2), namely, the *simultaneous realization of norms guided by the qualifying norm* for a particular area of human life. A third principle relates to the fact that systems methodology usually concerns human activity in which a diversity of human actors are involved. So *disclosure results from a multi-actor process* in which experts bear the responsibility to build a framework of cooperative responsibility for human action (section 3.3). Fourth, in building such a common framework the expert needs a *critical awareness of the social-cultural context* (section 3.4).

3.1   Disclosure and intrinsic normativity

Human culturally formative activity differs radically from events and processes in the realm of nature, such as the formation of rocks, the eruption of a volcano, or the gathering of honey by bees. Human formative activity is characteristically based on a free design. While processes and events in the realm of nature are subject to fixed laws, culture is a result of free human activity, which regularly brings new things to light. Thus again and again, human culturally formative activity

---

[5] See for similar views of the disclosive process of modern society part IV of Goudzwaard (1997).

that produces new things discloses new possibilities.[6] 'Disclosure' suggests that there is a connection between free forming and conditions that are constitutive for human activity. Actual creative, formative activity is not an imposing of man's will on the surrounding world but a sensitive responding to the world. In the response, specific circumstances conditioned by time and space must be taken into account. Sometimes the circumstances are a stimulus, but not infrequently they are an inhibitor or even an obstacle to human, renewing action. The notion of creative action means that people find ways to evade or escape such circumstances and in this way to create room for new possibilities.

Yet beyond the situational, the context of action, there must also be sensitivity to a deeper order of the world. Consider in this regard the artisan, who with his tools, practical knowledge and skills takes physical materials in hand. He knows from experience the limitations of his materials, which are inherent in its properties. And something similar holds at the theoretical level for the design engineer. In designing a bridge, for example, an engineer must take into account the mechanical properties of the materials to be used.

Creative and renewing activity accordingly manifests two aspects: it breaks through inhibiting circumstances but at the same time accepts structural limitations. Thus the popular slogan about the limitless possibilities of modern technology is based on a naïve and serious delusion about what actually occurs in technology. The technical ingenuity that produces new possibilities has nothing to do with escaping properties of matter or the laws that govern it. As Francis Bacon said, nature can only be conquered through obedience to it: *Natura non nisi parendo vincitur (Novum Organum* I 3). The properties of matter and the laws of nature form a boundary for human action. Man is tied and connected to them and restricted by them. What is indeed possible, however, is a creative human response to this order. It is precisely because creative and culturally formative activity bears the character of a response that there is room for input from the human side and with it also for the new. The invention of the airplane, for example, does not mean that man has succeeded in undoing the law of gravity. To the contrary, without gravity something like flight would be impossible. Flight is possible because

---

[6] The view on human beings as disclosers has recently been elaborated by Spinosa, Flores and Dreyfus (1997) in the line of phenomenology. The idea of disclosure in this paper is related to Dooyeweerdian philosophy.

gravity is compensated by another force, in this case the lift exerted on the wings.

What has been said here thus far about the technological forming of the physical, material world will doubtless find widespread support. And it may be asked whether that is so for technological activity in the broader sense? What does creative activity mean when the 'human factor' is at stake? When people take in hand social relations and structures, do their activities in this area too not bear the character of a response, thus implying that there is a significant degree of freedom for human form-giving with respect to social and cultural reality but also a being 'restricted' by or 'tied' to structural conditions that are given beforehand? But what, in that case, does 'restriction' mean? Are the boundaries here of the same sort one encounters in the physical world? More such basic questions could be framed in this way. From the perspective of 'disclosive systems thinking' there is indeed such a limiting tie. For the sake of clarification, let us consider briefly once again the situational and the structural.

In the first place the situational: and thus the fact, noted above, that culturally formative activity occurs within a historically conditioned situation. This means that there is a certain tie to circumstances that are given, and to the established power of tradition vested in them. When intervening creatively in social relations and structures one may accordingly not ignore the circumstances, which have been conditioned by time and location, which is to say by tradition, culture and history. In general there will be a state of tension between tradition and the forces of renewal. Yet such tension is healthy and should not be abolished. The negation of history and tradition can easily lead to the sort of wholesale destruction that results in the loss of what is valuable from the past. Preserving the tension in question while engaging in creative and renewing activity can contribute to tying the present and the past together. Dooyeweerd (1979:71) spoke in this connection of the *norm of historical continuity*. Human creativity is tied to this norm, which means that man "cannot create in the true sense of the word." Man is tied to the order of the reality that is given him. But it is precisely this tie that also determines our relative freedom and the possibility of forming culture. "If the past were completely destroyed, man could not create a real culture."

Yet as in the case of the physical world, one may also not ignore the structural when intervening in social reality. To the extent that disclosive activity bears upon the normative sphere of human life, such activity comprises a response to an intrinsic normativity that is given with that sphere itself. According to the nature of the normativity, however, the tie to it is of a different character than that of the tie to natural laws. People are simply bound by natural laws, but they can ignore normative laws (which is of course not the same as abolishing them).

3.2   Simultaneous realization of norms led by a qualifying norm[7]

If human formative activity must be looked upon as disclosive, responsive acting ('disclosure by response') upon an intrinsic normativity that is given with reality, then a question arises concerning the character of this normativity and of the human response connected with it. To facilitate answering this question it will be useful to introduce first, briefly and schematically, the ontology we employ.

In the first place we make a fundamental ontological distinction between law and (created) reality. This distinction, which was implicit in 3.1, means that the great diversity of matters that appear in the human experience of reality - stones, water, air, plants, animals, people, social relationships and structures, man-made material artifacts, processes, human relations and actions, and ever so much more - corresponds with an order that obtains for reality. Law and reality are not two separated 'realms'; rather, they have a bearing on each other. In the being of things as they are, the law that obtains for such things manifests itself. The capital ontological distinction between law and reality implies that 'disclosive systems thinking' is founded on an uncommon view of the relation of God to reality. To make this clear we may pause for a moment to consider Ackoff's position.

In a number of his publications Ackoff (1976, 1981, 1994) deals explicitly with the ontology that underlies his systems approach. He contrasts the systems approach as synthetic scientific method to the analytic method of the 'Machine Age'. In the 'Machine Age' things are broken down into their component parts, and the whole is approached from the vantage point of the parts. In the synthetic method of the 'Systems Age' in contrast precisely the opposite is the case: the whole is not approached from the vantage point of the parts, but the parts are understood from the perspective of the whole. Behind these two methods

---

[7] The expression 'simultaneous realization of norms' is adopted from Goudzwaard (1979) and Van der Kooy (1953).

Ackoff discerns two differing views of the relation of God to reality. Characteristic for the 'Machine Age' is the deistic view in which God is regarded as the Creator of a machine which runs according to fixed laws (God as watchmaker). In the beginning God was involved with reality, which was subsequently consigned, however, to the operation of the fixed laws of nature. The machine was set in motion and God then withdrew from his work. Ackoff rejects this ontological view of the relation of God to reality because in it there is no room for human freedom and the possibility of influencing the course of events. The future is determined and there is no room for the kind of systematic intervention whereby man might influence his future in any real way at all. Instead of our determining our future ourselves, our future is determined for us. All is laid down in the iron laws of the machine.

To create room for freedom Ackoff rejects the deistic distinction between God and the world. The systems approach as Ackoff would deploy it proceeds from a pantheistic view in which God and the world coincide as the largest, all-embracing whole. While the 'Machine Age' and deism personify God as the Creator-God, God loses this personal and individual character in the 'Systems Age'. "This God," Ackoff (1981:19) asserts, "is very different from the Machine-Age God who was conceptualized as an individual who had created the universe. God-as-the-whole cannot be individualized or personified, and cannot be thought of as the creator. To do so would make no more sense than to speak of man as creator of his organs. In this holistic view of things man is taken as part of God just as his heart is taken as a part of man." The whole of reality is thus God, and God is the whole in which man-and-world are incorporated.

Now, various critical questions can be posed with respect to all this, but I shall restrict my comments to Ackoff's ontology and leave undiscussed here its consequences for his scientific methodology of planning. I can concur with him that a deistic view of reality can not do justice to human freedom; in this view man is just a little cog in a field of forces entirely beyond his influence. Yet how does freedom fare in Ackoff's pantheistic view of reality in which the distinction between God and reality has vanished? When the boundary between God and created reality is obliterated, does not man come to occupy a remarkable dual position as master and slave of a cosmic whole in which he is incorporated? Is human freedom any less problematical in Ackoff than it is in the deistic position he criticizes? So the question arises: is there an alternative to the two contrasting views described by Ackoff that can resolve this problem?

It is not my intention to explore here the various views that have been advanced concerning the relation between God and reality. One thing however will have become clear from what has been said thus far: it does not follow from the deistic conception's failure to satisfy that a pantheistic view must perforce be accepted. Essential for 'disclosive systems thinking' is its maintenance of the distinction between God and created reality. The connection between God and the world is regarded differently however than in deism. It is given with the creation ordinance of God. Instead of proceeding from the distinction between God and created reality, 'disclosive systems thinking' proceeds from a distinction between God, law and reality. There is also a second difference from deism (and also from the pantheistic view proposed by Ackoff). According to 'disclosive systems thinking' man, including his thought, is incorporated in created reality which is subject to the law. And this means that from this point of view the relation between God and the world cannot be determined by theoretical thought, man does not have a 'God's eye view on reality'. The law forms a boundary for human action, including human thought. The law must however not be viewed as a sort of model or scheme that only functions as a blueprint for the world. No, the law of God expresses a living and continuous relation between God and reality; through the law created reality is connected with God in its origin, existence and end. That means in other words that created reality has a referential character. One could speak here of the responsive character of reality (Geertsema, 1992, 1993). The law restricts man but at the same time creates room for properly human responsibility. Man is not a cog in the cosmic machine (deism) but in his activities contributes creatively to the progressive disclosure and development of reality.

Distinctive of 'disclosive systems thinking' is thus the idea of the law and the correlation between law and reality. Both the law itself and the things, which correspond with the law, manifest a rich diversity. In order to grasp this diversity it is necessary to introduce a second capital ontological distinction, namely, that between entities and modalities. First a word about entities.

By entities we mean the individual wholes that present themselves in concrete experience, such as a tree, a horse, a building, a book, a university, a firm. Instead of entities one could speak here of systems, albeit it must be kept in mind that not all systems have the character of an entity or individual whole. Sometimes there is an interweaving of different wholes. The system is in that case an interwoven whole. An example of the latter would be that of a bird and its nest. Both individualities retain their own structure in this unit or interweaving. In other words, the bird and the nest cannot be regarded as parts of a whole. The whole-part relation is only applicable, strictly speaking, when the whole and the part have the same character. Think of a tree as a whole, for example, and of the trunk and branches as its parts.

We have seen that a distinction can be made between various sorts of systems. There are entities, individual wholes with their own proper structure, and interwoven wholes made of different entities. Now, what determines the distinction between different entities? For example, what makes a tree as entity different from the bird that sits in its branches and from the traffic lights and mosque in its shadow? A first reaction to this question might be that the traffic lights and the mosque are man-made entities while the tree and the bird in its branches were produced by the natural world. The distinction 'man-made' or 'produced by the natural world' explains something but certainly not everything. For on what basis would one go on to distinguish between all that falls within one of these categories?

To move forward here it is helpful to consider the distinction drawn earlier between entities and modalities. Namely, each entity functions in a diversity of modalities or modes of being, which are aspects of one and the same entity. And these aspects or modalities may be put in a particular order. Some aspects presuppose others. Thus the physical modality is possible without the biotic, as in the case of a stone, but the reverse is not so. And the psychic or sensitive cannot exist apart from the biotic - consider the bird in the tree - while the reverse is possible and the biotic can exist without the psychosensitive - consider the tree. Upon comparing the bird and the tree, we can say that both function in all the aspects mentioned thus far, physical, biotic and psychosensitive, with this difference, that the bird functions subjectively and the tree objectively in the psychosensitive aspect, inasmuch as the tree can be the object of the subjective feelings of the animal.

Thus when referring to modalities we have in mind the aspects we can distinguish in entities and in our human experience of entities. Of a book that we take in our hands we can say, for example, that is beautifully made, that we enjoy its attractive design, the color of the paper, the elegant type, and much more (the aesthetic aspect). If subsequently we look at the price (the economic aspect), we may exclaim spontaneously that this book is beautiful, to be sure, but terribly dear. And we may be inclined to put it aside for this reason. Yet at the same time we may realize that this book is precisely what our friend would want for her birthday (the ethical aspect). We alluded here in passing to three aspects of reality that can be connected with books. These aspects, which are normative in character, stood out as we considered whether or not to purchase a book; the concrete act of purchasing (or not purchasing) takes form in a process of a simultaneous realization of norms. That is so not only for this example; all responsible actions must in fact be understood as simultaneous realizations of norms, whereby no aspect may be neglected. Yet with that, still not everything has been said about responsible action. For the question arises: how must the norms for the diverse aspects be weighed in their mutual interrelationship? What leads the process of simultaneous realization of norms? The answer to this question is given by the distinctive character of the action - in our example, the making of a purchase. The simultaneous realization of norms in an action must thus be led by the qualifying norm[8] for that action.

3.3   Disclosure as multi-actor activity

In 3.1 I discussed how culturally formative activity is responsive in character and argued that responsiveness to the normativity that is given with reality results in disclosure of new possibilities. In 3.2 it was shown that such disclosure evinces a diversity of norms that must be taken into account. Disclosive and responsive acting is in need of a simultaneous realization of norms led by a qualifying norm. Now I want to consider the actors involved in the action. In general, many actors are involved and they hold a wide variety of positions and responsibilities. That is certainly so when the object of formative activity is comprised of people in their social interconnections and relations. Then 'disclosure by response' occurs as a multi-actor activity. The intrinsic normativity that is given with a specific normative sphere of life concerns not only the nature of the activities within the sphere upon which the formative actions bear but touches also the relations between the actors engaged in them. In other words, the systems ethics that underlies 'disclosive systems thinking' as a systems methodology - and which I have called earlier an 'ethics of disclosure' - also addresses the question of the normative relation between actors, and particularly the question of the role of the scientific expert in relation to the other agents.

---

[8] A practical application of the idea of qualifying norm is discussed in chapter 3.

Different systems methodologies may differ substantially respecting their view of the scientific expert's role and responsibility in altering reality (cf. Ulrich, 1991). Here a brief indication must suffice. In 'hard systems thinking' the perspective of the scientific expert is central. The plan s/he develops is adopted and implemented by the client; the expert's plan and its implementation are phases in the process of change. Underlying this approach is the notion that the expert has the capacity to assess the situation from the outside in a completely objective manner and to determine from the perspective of an outsider what is good for those concerned.

In 'soft systems thinking' the role of the expert as an outsider is relativized. The expert takes into account that the position of other actors besides the client is at stake. The matter is now not just one of proposing a plan the implementation of which is to be left to a primary client. Now one of the most important problems is to determine which actors are concerned with the change in one way or another. In other words, how is the expert to delimit the system that is to be the focus of change? Where are the boundaries of the system? In 'soft systems thinking' the expert remains an outsider but as an outsider responsible for identifying all those who may have an interest at stake and for seeing to it that they are all given a place in a strategy for change which enables their outlook to be taken into consideration. According to the methodology of 'soft systems thinking' the expert proposes not so much a plan as a strategy for change.

The relativization of the independent position of the expert as outsider in 'soft systems thinking' is carried further in 'critical systems thinking'. The position of the independent outside expert is in fact abandoned in favor of the perspective of the expert as a committed participant engaged in the concrete situation. It is the expert's task to instigate a discussion between all the participants concerning the desired changes. The starting point is fairness amongst all the participants having something to contribute, such that by means of a critical discussion a consensus can be formed about the situation and what is to be done about it.

Against the background of these views of the role of the expert in 'hard', 'soft' and 'critical systems thinking' it is now possible to clarify the role of the expert in 'disclosive systems thinking'. 'Disclosive systems thinking' shares with 'soft' and 'critical systems thinking' relativization of the idea of role of the expert as an objective outsider. The question of delimiting the system, which is to say the question of which actors are to be considered 'clients' of the expert, is a question to which 'disclosive systems thinking' also seeks an answer. Of guiding significance here is an acknowledgment of the personal responsibility borne by each of the actors in the collective action situation investigated by the expert, together with the intrinsic normative structure of the situation. While 'critical systems thinking' places the relation between the actors against the background of power, inequality of power, and breaking through such inequality, the point of departure in 'disclosive systems thinking' is that justice be done to the responsibility of each of the actors.

It cannot be denied that power and inequality of power are phenomena that lead to people being shortchanged in action situations. But the abolition of inequality will not necessarily lead to the emancipation of people that 'critical systems thinking' desires if it remains unclear what the norm for human freedom and responsibility is. Abolition of differences of power will not lead directly and automatically to responsible action. This is the matter to which 'disclosive systems thinking' seeks to provide an answer.

3.4   Critical awareness of the social-cultural context

In shaping a common context for the collective action situation in which each of the actors can discharge his or her personal responsibility, the expert must take into account the social-cultural context of the action situation. First there is the larger context of the global technological society with its characteristic developmental tendencies and features (cf. Strijbos, 1997). But there is also the local context of the problem situation, with all its specific features and tendencies, in which action must occur. The context, whatever it may be, always influences to a greater or lesser degree the actions of the actors. The expert must always remain well aware that these contextual influences can move actors to engage in actions inconsistent with the common framework of responsibility the expert is seeking to shape. The problem situation must be analyzed from the standpoint of a *critical awareness of the social-cultural context*, and a strategy for action must then be developed for approaching the situation in a normative way in order to bring it to further disclosure.

The pressure to adapt that is exerted by the global and local contexts can be enormous. The normative action principle of 'critical awareness' means that one must take this into account. So far the term 'critical awareness' as used here is in line with 'critical systems thinking'. But this principle also means that one is critically aware of the fact that norms do not have the status of purely human constructions, which means that the intrinsic normative structure of reality always obtains and never (as we saw at the close of 3.1) allows itself to be negated. In the perspective of 'disclosive systems thinking' there are therefore always possibilities for a situation to disclose itself further, even if the developmental tendencies of the global and local contexts seem diametrically opposed to disclosure of the situation concerned in a normative direction. 'Critical awareness' leads thus to insight into the dominant direction of development of the social-cultural environment in which people live and act, the situational, but also to an understanding that this direction remains tied to what in 3.1 is called a 'deeper order of the world', the structural.[9]

4. CONCLUDING REMARKS

The normative principle of 'critical awareness' that gives direction to 'disclosive systems thinking' as a systems methodology is also applicable to the further development of this methodology itself. 'Critical awareness' with respect to systems methodology brings awareness that the dominant direction of development in this area of science has thus far been conditioned by the idea of autonomy and at the same time that this direction of development remains tied to the intrinsic normative structure of this field of reality, which is to say systems methodology as a scientific method for intervening in and altering the world.

This perspective has important implications for the critical position of 'disclosive systems thinking' in the field of systems methodology. Although autonomy and heteronomy are diametrically opposed and mutually exclusive choices for the fundamental direction of thought, there remains a possibility for open communication. From the standpoint of 'disclosive systems thinking' the idea of autonomy as the basic choice of direction is subjected to criticism, but at the same time this standpoint offers openness to other systems methodologies. It acknowledges that in these other directions of thought, systems methodology as a field of science has undergone further disclosure. The path of development from 'hard' via 'soft' to 'critical' systems thinking and beyond as traced in 2.2 is itself to be understood as a continuing process of disclosure to which 'disclosive systems thinking' as the latest development on the field of systems methodology seeks to make a contribution. The opposition in principle with respect to the choice of direction (autonomy versus

---

[9] See Wolters (1986) for the distinction as made here between structure and direction.

heteronomy) does not negate the fact that in the concrete elaboration of a methodology elements may be developed that can be adopted.

With that we have arrived at the current debate in systems thinking about methodological pluralism and complementarism (cf. Mingers and Gill, 1997). However, this theme lies beyond the scope of this book and forms a subject to be addressed more fully in some future publication.

Part V:
Critical Reflections

Chapter 15

REFLECTIONS ON THE CPTS MODEL OF INTERDISCIPLINARITY

Gerald Midgley

1.   INTRODUCTION

In this chapter, I adopt the role of 'critical friend' to the CPTS research programme[1]. I believe the CPTS model of interdisciplinarity has some significant strengths, and also some potential weaknesses that the researchers taking it forward might wish to address. Most of my critique refers to the introductory chapter of this book, as this offers the grounding for the rest of the CPTS research programme. However, my focus on the introduction should not be taken as a sign that I regard other contributions to the book as less significant - it is just that the basic CPTS model of interdisciplinarity is my primary concern.

Over the coming pages I will first of all highlight what I see as the strengths of the CPTS model, focusing in particular on the value of the systems approach embodied in it, and its potential applicability to technologies beyond information systems (the practical focus of most CPTS authors to date). I will then offer two critiques. The first points to a gap in the model: the omission of ecological systems as an aspect of analysis. The second critique raises some questions about the nature of the links between research at the levels of the artifact and directional perspectives. I suggest that, when there are significant disagreements on the ethics of a technology, to the extent that some researchers wish to

---

[1] [The CPTS, the Centre for Philosophy, Technology and Social Systems, and its research programme from which this book has emerged, are described in the Preface. What Midgley calls the CPTS model is that depicted in Fig. 1 of chapter 1. Eds.]

prevent its development and others wish to press ahead, we have to ask whether and how interdisciplinary co-operation should proceed.

2.   THE STRENGTHS OF THE CPTS MODEL

In my view, the CPTS model of interdisciplinarity has several important strengths: it is explicit about its theoretical underpinnings; is inclusive of ethical debates; takes a useful systems approach to understanding the relationships between fields of inquiry; is potentially applicable to a broad range of technologies; and can enable the incorporation of many more disciplines than are currently included in the CPTS research programme. I discuss each of these strengths in turn below.

2.1   The value of explicit theory

The first strength is that there is an explicit theoretical rationale for the focus on basic technologies, technological artifacts, socio-technical systems, human practices and directional perspectives as the principle concerns flowing into interdisciplinary engagements. As Strijbos and Basden (chapter 1) make clear, these categories are derived from the philosophy of Dooyeweerd (1955). Although I am not in complete accord with Dooyeweerdian thought, I nevertheless appreciate that there is a coherent set of ideas lying behind the CPTS model. This is important because it takes us a step beyond models that are simply born out of strategic alliances between researchers from two or more disciplines who happen to share common interests. While alliances like these can be useful for pursuing focused projects with particular purposes, it is difficult for them to give rise to more general models of interdisciplinarity unless there is a focus on providing some theory that explains why the model might have utility beyond the immediate local circumstances in which it was generated.

2.2   The incorporation of ethical considerations

In addition to being explicit about theory, the CPTS model is inclusive of ethical considerations under the heading of 'directional perspectives'. This is important because there is a tendency in modern societies for ethical issues (about which ends to pursue and why) to be separated from technical ones (how to implement the ends that have already been pre-determined) (Habermas, 1984, 1987). Even some supposedly participative approaches to information technology planning give people scope to debate means (ways to implement technologies) but not ends (the missions of their organisations that give rise to desires for technological solutions) (Willmott, 1995). By incorporating the research domain of 'directional perspectives' into the CPTS model of interdisciplinarity, and by making it clear that these can *frame* debates about technology (as well as being impacted by technological innovations themselves), it becomes much more difficult to marginalise ethical concerns than might be the case if the human dimensions in the model were restricted to socio-technical systems and human practices. Clearly, the strong inclusion of ethical considerations comes about because of the theoretical influence of Dooyeweerd (1955), but it makes the model equally useful from a critical theory standpoint (e.g., Habermas, 1984, 1987) or a critical systems perspective (e.g., Ulrich, 1983; Jackson, 1991; Gregory, 1992; Oliga, 1996; Midgley, 2000). For most writers on critical systems thinking, ethical reflection and dialogue are essential aspects of inquiry (interdisciplinary or otherwise).

2.3 The systems approach

The CPTS model also offers a strong *systemic* conceptualisation of the relationships between the various kinds of research that flow into interdisciplinary practice. Strijbos and Basden (chapter 1) focus on the *integration* of ideas across the levels of basic technologies, technological artifacts, socio-technical systems, human practices and directional perspectives. Here, they draw upon Boden's (1999) understanding of integration (one discipline learning from another), although there is actually a long tradition of integrative research going back to some of the earliest work in systems science (see, for example, Bogdanov, 1913-17; von Bertalanffy, 1956; Boulding, 1956; Miller, 1978; Troncale, 1985; Bailey, 2001; Midgley, 2001). Many authors have tried to transcend the limitations imposed on inquiry by seemingly arbitrary disciplinary boundaries. While some of these (e.g., von Bertalanffy, 1956) have viewed integration as the generation of a new 'general system theory' to complement or even replace the old disciplinary ones (Boden, 1999, is critical of this), others take a different view. It is especially interesting to read Boulding (1956), who offers a 'skeleton of science' that is structured into similar levels as the CPTS model, and Boulding even recognises the relevance of spirituality - although there are actually more levels in Boulding's framework (and a tighter hierarchical relationship between them[2]) because his purpose is to provide a model for use across the disciplinary sciences, not just within the field of technology.

2.4   Applicability to a broad range of technologies

Although the CPTS interdisciplinary research community has taken information systems as its first application domain, Strijbos and Basden (chapter 1) are explicit that their desire is to generate ideas that can be of use to research a wider set of technologies. I have therefore decided to test the wider applicability of the CPTS model through two simple 'thought experiments'. I have taken two technologies - workplace drug testing and genetically modified organisms (GMOs) of use in food production - to see whether the levels of analysis in the CPTS model are able to account for the various different issues that I am aware are being researched in these areas. I am not a specialist in either of these fields, yet I have taken an interest in some of the issues associated with them. Each is discussed in turn below, starting with workplace drug testing.

The basic technologies of workplace drug testing are chemical markers that indicate the presence of illicit drugs in urine and saliva samples. These chemical markers are the basis for the production of testing kits (artifacts). The kits are deployed within socio-technical systems: organisations wishing to test their employees in order to improve safety in the workplace (drug testing is generally introduced in relation to safety-critical occupations, although some employers use it more widely). Various human practices may be impacted, including personnel selection (drugs testing can become part of the recruitment process), counselling for people with drug and alcohol problems (many testing regimes are introduced alongside rehabilitation programmes) and drug-taking behaviour (people may stop taking drugs, moderate their use, or shift to

---

[2] Boulding (1956) proposes a tight hierarchy, with simpler, smaller sub-systems being the 'building blocks' for the emergence of more complex, larger-scale systems. While there is a *general* movement from small to large in Strijbos and Basden's (chapter 1) list of basic technologies, technological artifacts, socio-technical systems, human practices and directional perspectives, I know these authors are aware that a strict hierarchical representation is problematic. The problems become particularly evident when you look at the relationship between socio-technical systems and human practices. A socio-technical system can be as small as a department within an organisation or as large as the global economy. Therefore, the relationship between socio-technical systems and human practices cannot be described simply as a class of systems (socio-technical ones) within a wider human environment: some socio-technical systems may *contain* human practices, and other human practices will be outside, and mutually influencing, those systems. The exact relationship between socio-technical systems and human practices therefore needs to be defined in a locally meaningful way within each interdisciplinary research project.

drugs that are less easy to identify in a urine or saliva sample). Finally, at the level of directional perspectives, various ethical issues are relevant: e.g., those surrounding the tension between public safety and personal freedom. It seems to me that the CPTS model can capture all the main concerns of researchers looking at workplace drug testing, and it reveals substantial scope for interdisciplinary engagement.

Next we can look at GMOs. At the level of basic technologies, the functions of various genes have been identified, and new genetic combinations with desired properties have been developed. At the level of the artifact, crops are produced (e.g., genetically modified, disease resistant maize plants) using the results of the basic genetic research. These are then deployed within socio-technical farming systems, and these in turn interact with larger systems, including those associated with retail and international trade. Human practices of farming and eating are affected, as are political practices (e.g., there may be an increase in direct action protests). Finally, at the level of directional perspectives, the ethics of genetic modification are debated in research publications, the media and amongst ordinary citizens.

In the GMO example, I suggest that *most* (but not all) of the relevant research themes are accounted for by the CPTS model (I say 'most' because ecosystem research is not explicitly included, and I'll pick up on this later in the chapter). Most importantly, the need to link together research at the various levels becomes quite apparent once we explore the connections between them. My own view is that the basic/artifact research on GMOs is still, by and large, overly disconnected from ethical research, despite the fact that many scientific authorities now recognise that the GMO issue (together with some other issues debated in the latter half of the 20th Century) has brought the whole credibility of ethically-disconnected science into question (e.g., ESRC Global Environmental Change Programme, 1999).

Based on the two examples above, and the CPTS research on information systems presented elsewhere in this book, I believe it is reasonable to conclude that the CPTS model of interdisciplinarity may well be useful for research across a range of technologies (but with some caveats, to be explained shortly).

2.5   The incorporation of a wide range of disciplines

A final strength of the CPTS model is that it has the potential to incorporate a wide range of disciplinary perspectives from the sciences and humanities. In relation to information systems, the various chapters in this book demonstrate the inclusion of computer engineers, information systems practitioners, management scientists, systems thinkers and philosophers within the CPTS interdisciplinary network. However, this is a relatively limited range of disciplines in comparison with those that might need to be involved in interdisciplinary research on workplace drug testing (biochemists, manufacturing technologists, occupational health specialists, organisational analysts, economists, psychologists, psychiatrists, social workers, sociologists, policy analysts, systems thinkers and philosophers) or GMOs (biologists, agricultural scientists, economists, political scientists, sociologists, ecologists, systems thinkers, philosophers and theologians). The disciplines in brackets are just my own suggestions for inclusion - the potential scope is no doubt wider.

3.   CRITIQUES OF THE CPTS MODEL

Having highlighted what I see as the main strengths of the CPTS model of interdisciplinarity, it is now time to look at two potential weaknesses: the absence of an explicit focus on ecosystems, and what appears to be the assumption that scientific research into basic technologies and artifacts can sit harmoniously alongside philosophical research on directional perspectives, even when philosophers are advocating the abandonment of the technologies in question. I deal with each of these in turn below.

3.1   Ecosystems research

The 'thought experiment' on GMOs that I briefly described above highlights a missing level in the CPTS model: the level of ecosystems. Of course, one could argue that ecosystems research needs to be conducted as part of the existing foci of the model: at the levels of the artifact (where ecological impacts of GMOs might be assessed), the socio-technical system (which people might claim includes ecological elements alongside the technical and social ones) and directional perspectives (where ecological arguments could be used to support either pro- or anti-GMO positions). However, it is *always* the case that the ecological, ethical, social and technical levels are all relevant to one another - it is precisely the point of the CPTS model to make explicit the relationships between the various levels. Therefore, to make the ecological implicit in the technical, ethical or social is to accept an aspect of the reductionist rationality that the CPTS model has been designed to challenge.

Worse than this, I suggest that the marginalisation of ecological concerns is systematically prevalent in Western political (and also many academic) discourses and practices, although thankfully less so than just one generation previously. There is therefore a danger that, left unaltered, the CPTS model will act to reinforce this marginalisation. I say that the marginalisation of ecological concerns is *systematically* prevalent in Western discourses and practices because I believe that marginalisation processes are far from random. Elsewhere (Midgley, 1994), I have written about this at length. Here I shall simply point out that the marginalisation of environmental issues has resulted from the dominance, over several hundred years, of anthropocentrism (seeing humankind as the centre of things, somehow disconnected from our environment) - and Western philosophy has not been exempt from making anthropocentric assumptions. Even some systems thinkers (let alone philosophers) root the origins of rationality in either the individual human mind alone (following Kant, 1787) or linguistic communities (following Habermas, 1984, 1987), thereby ignoring Bateson's (1972) insight that both mental and social phenomena interact with ecological systems. From Bateson's (1972) perspective, rationality can be seen as a product of the wider systems we participate in - not a product of human beings or communities in isolation (also see Midgley, 2002).

If the proponents of the CPTS model want to take this point seriously, they will be faced with a dilemma: either remain faithful to their original translation of Dooyeweerdian philosophy into a framework for interdisciplinarity, thereby preserving the marginalisation of ecosystems research, or further develop the model to incorporate the ecosystems focus. Without conducting some new research, I am unsure whether or not this will necessitate revising some of the original Dooyeweerdian concepts, but in my view the whole issue is worth looking into. As I see it, exploring the ecological impacts of technologies (at local, regional and global levels) is a pressing priority, and we marginalise this at our peril.

3.2 Dealing with conflicts over normative beliefs

My second critique of the CPTS model comes from asking the question, "what if some researchers wish to prevent the development of a technology?" It seems to me that the CPTS model already presupposes the existence of a given technology (such as information systems), and the task of the interdisciplinary research community is to bring their various perspectives to bear on it, supporting each other in making everybody's work more systemic. This is certainly a laudable aim, but what should we do when a technology is at a conceptual or early developmental stage and normative explorations at the level of directional perspectives lead to a conclusion that it is illegitimate? In such circumstances, will philosophers of technology (or others engaged in research on ethics) be expected to co-operate with those whose mission is to bring the 'illegitimate' technology to fruition?

A rejoinder to this question from an advocate of the CPTS model might be that this is *exactly* what needs to happen: without interdisciplinary engagement there will be no systemic thinking about the technology and therefore no chance to affect its development. My problem with this answer is that it is a little naïve with respect to the power relations that surround the production and deployment of technologies - most technologies being produced by companies who make substantial investments in research and development. While they expect some ideas to fail, they also expect enough to succeed to yield a return for their shareholders. These companies therefore have significant vested interests, and the scientists working for them are rarely immune to commercial pressures: in many research and development divisions, the continued employment of scientists depends on the results they achieve. There is therefore an incentive for people working at the levels of basic technologies and artifacts to draw narrow boundaries around their research and exclude collaboration with people bringing them the very worst kind of 'bad news' (Gouldner, 1975) - that their new idea might, from some points of view, be considered completely illegitimate.

Again there might be a rejoinder from an advocate of the CPTS model. Surely closing off to this bad news is not really in the self-interest of a company developing a new technology. Doesn't *enlightened* self-interest dictate that the company should be aware of potential problems with the technology so that they can address them in advance of a commercially damaging crisis? It would be nice to believe that, if companies can be persuaded of the utility of a systems approach and the value of dialogue, then it will be worthwhile for philosophers of technology (and others with an interest in ethics) to engage with those developing a seemingly 'illegitimate' product - and on occasions it may be the case that this engagement will be meaningful.

However I suspect that, in the majority of situations, the volatile mixture of commercial self-interest, the desire for secrecy so that the company can gain competitive advantage over others in the same market, and fear and distrust of people with radically different perspectives will either prevent engagement altogether, or will limit this engagement to a tokenistic recognition of other points of view without there being any real prospect for changing the technology in question (we can call this tokenistic attitude 'pseudo-dialogue'). In the case of engagement that is completely blocked, the philosophers of technology (and others with ethical concerns) will know where they stand: they will be better off working independently, or through alliances with other stakeholders, to make their case in various civil society fora. It is the tokenistic form of engagement that is more worrying: it is conceivable that the CPTS model might be used to demonstrate a coherent logic of engagement, thereby allowing ethicists to be 'captured' (or even duped) by those who have no real intention of reflecting meaningfully on their chosen path for action.

The issue is therefore whether use of the CPTS model of interdisciplinarity may, in situations where there is a strong normative conflict, result in a bias towards the values of the developers of a technology, with ethicists getting unwittingly tied up in pseudo-dialogues with their opponents. Anyone who is sceptical about my critique might ask themselves how often scientists with a nascent technology, employed by a company which has invested in its development, knowingly abandon that technology after hearing the arguments of philosophers. I would love to be proven wrong, but I suspect that this is a very rare occurrence indeed.

If the proponents of the CPTS model want to take this point seriously, I suggest it should result, not in the abandonment of the model (it has some significant strengths, and represents an ideal of good practice), but in further critical reflections on when and how it should be used. If there is a chance of the CPTS model being co-opted to promote pseudo-dialogue rather than meaningful engagement, then social researchers might need to think about how they explore situations characterised by value conflicts and power relationships prior to, alongside of, and/or instead of engaging with technology development. For this purpose, some of the literature on critical systems thinking (e.g., Ulrich, 1983, 2001a, 2001b) and systemic intervention (e.g., Midgley, 2000; Córdoba and Midgley, 2003) may be useful, as writers in these areas have been working with questions around values, boundaries, power and marginalisation processes for some time.

4. CONCLUSIONS

In this chapter, I have sought to reflect on the strengths and weaknesses of the CPTS model of interdisciplinarity so as to support its further development. In my view, there are some significant strengths to the model that make it *worth* developing. In particular, it is explicit about its theoretical underpinnings; is inclusive of ethical debates; proposes systemic relationships between fields of inquiry; is potentially applicable to a broad range of technologies; and can enable the incorporation of many more disciplines than are currently included in the CPTS research programme.

However, there are also some potential weaknesses that only come to the fore once we think of the model in relation to technologies other than those to which it has already been applied. My reflections on the GMO issue have raised a question about where ecosystems research might fit. I suggest that a new 'level' (ecological systems) is needed in the CPTS model, and further work would be useful to see whether this adaptation will necessitate any rethink of the philosophy underpinning the CPTS research programme. The controversial nature of the GMO issue also raises a question about how those developing a technology and those opposing its development could realistically be expected to collaborate on interdisciplinary research. As I see it, the worst-case scenario is not a breakdown of dialogue (then people know where they stand), but co-option of the CPTS model by vested interests to enable a *pseudo*-dialogue that effectively neutralises the perspectives of those arguing that a technology is illegitimate. To avoid this kind of scenario, proponents of the CPTS model may be able to learn more about how to explore situations characterised by value conflicts from people in neighbouring research communities - for example, those engaged in critical systems thinking and systemic intervention. This paper can be seen as a first step in opening a dialogue between people involved in the CPTS research programme and systemic intervention practitioners.

Chapter 16

TECHNOLOGY AND SYSTEMS - BUT WHAT ABOUT THE HUMANITIES?

Carl Mitcham

> *Apologia: The following is meant as a provocative epilogue more than a fundamental challenge to a project with whose thesis I am in fundamental sympathy.*

1. INTRODUCTION

Among many inhabitants of advanced industrial society there has been for more than two centuries a certain ambivalence about technology. Such ambivalence began to find sporadic expression in the 18th century Romantic criticisms of science and technology, then took more expansive forms in 19th century fiction and philosophy. Considering only the fiction, one particularly vivid representative of this genre was Mark Twain's novel *A Connecticut Yankee in King Author's Court* (1889).

Connecticut Yankee is the first novel based on time travel. An engineer-inventor, Hank Morgan, gets in a fight in the mid-1800s in a Connecticut arms factory, is hit over the head, and recovers consciousness in the mid-500s Arthurian England. On one level Morgan's adventures aim to undercut the romances of Sir Walter Scott (1771-1832) and to show the early medieval period for what it really was: solitary, poor, nasty, brutish, and under the sway of slavery and superstition. Morgan's triumphs in science, technical ingenuity, and science policy - he predicts an eclipse, manufactures soap and blasting powder, creates a patent office, establishes an engineering school, electrifies Camelot - are all designed to show the superiority of Yankee civilization. At the same time the narrative discloses how Morgan's faith in technology is structurally similar to Arthurian beliefs in Merlin's magic, and turning the Round Table into a stock exchange raises questions about the sacrifice of virtue to commercialism. In the final apocalyptic battle Morgan's new technology winds up killing thousands of people and Morgan is trapped in the stench produced by his own success. After his return to the 19th century he remembers with nostalgia his time in the 5th.

Technology in the 21st century is likewise often seen as both of great benefit, a uniquely human achievement, and of serious danger, a threat to other goods that are equally worthy of respect. Thomas Hughes' argument in *American Genesis* (1989) and *Human-Built World* (2004) that the great age of American inventiveness from 1870 to 1970 was a historico-cultural achievement comparable to that of classical Greece and Renaissance Italy remains ultimately unconvincing in the wake of the problems that ensued. The qualified skepticism that still cannot decisively reject such claims is manifested in the popular argument that whatever problems exist result not from the inventions themselves but from our uses of them: technology is just a neutral means that we must learn to use wisely.

One difficulty with this response is that the special powers of modern technology arise precisely from what the economic historian Karl Polanyi (1944, 1957) has described as a process of *disembedding*. Disembedding is a dual process that takes place on two levels. On the general or external level, disembedding separates science and technology from culture as a whole, including the mores and counter mores institutions that would otherwise circumscribe and guide their development and deployment. Scientific and technical facts are to be distinguished from religious, political, aesthetic, or even personal values - and thus to be autonomously explored. On particular or internal level, disembedding takes place within modern science and technology through a methodological simplification of cognitive or practical problems that separates them from their real-life complexities. Take the example of free body diagrams in engineering. Such diagrams "treat an object *as if* all forces were acting directly on its center of gravity, in order to model matter-force interactions" (Mitcham, 1994b:162). But this 'as if' model denotes an abstraction that *is not* the complex reality, since in fact many forces act on the surface of an object which, given certain shapes, may introduce small or large deviations from the model. Yet it is this analytic breaking down of problems, separating them off from their rich contextual manifolds in a process that has been both praised and stigmatized as reductionism, that enables science and technology to develop their powers of knowledge production and artifact construction to a degree unprecedented in human history.

To talk about the wise use of these powerful means - means that they exist as such precisely because they have been disembedded from any kind of contextual complexity, including the traditional complexities of cultural governance - is passing strange. For those who appreciate this anomaly the alternative has been to argue for some reintegration of science and its technologies into a larger cultural context, such as the hierarchy of system levels proposed by Strijbos and Basden in chapter 1.

2.   SYSTEMS DISTINCTIONS

As Strijbos (1988) and his colleagues have noted, one can distinguish different approaches to systems. The term 'system' comes from the Greek *sunistanai*, meaning to combine (which is itself a combination of *sun-*, with or together, and *histanai*, to cause to stand), by way of the late Latin *systema*. The Greek noun *sustema* refers to an organized whole, especially in the form of a government, but also of a musical composition. (The prefixes of 'system' and 'symphony' are the same.) The Latin transliteration is used to refer to such things as the universe as a whole or the collected articles of faith. From the 17th century in English the word can mean an organized body either of objects or of principles.

When the term began to be used in the mid-20th century it was employed by different theorists to promote the development of science and engineering methods at odds with the typically modern, analytic separation of problems into their component parts. Although there are indeed mechanical systems in which the whole is simply the sum of its parts, systems science and systems engineering sought to emphasize ways in which wholes, because of the complexities of the interactions of their parts, can become more than their sums. Leaders in this new approach staked out what became four interrelated traditions or schools. In the technoscientific school, Wiener (1948) and Von Bertalanffy (1968) worked on systems approaches in mechanics and biology. In the social sciences school, Parsons (1951) and Luhmann (1984) used systems thinking to reconceptualize functionalist sociology. In the management school, theorists from Beer (1959) to Senge (1990) sought to turn systems thinking to business and political advantage. In the philosophical school, Churchman (1968b), Laszlo (1972), and Ackoff (1974), among others, have promoted the systems approach as a new worldview. The rise of chaos and complexity theory in the 1980s and after could be described as another version of systems thinking that crosses scientific and philosophical boundaries.

Systems thinking has nevertheless played a paradoxical role in advancing scientific knowledge and technological power since the end of the Second World War. On the one hand, it has criticized reductionist science and engineering. Each of the four schools in its own way has promoted receptivity to and awareness of interconnectivity in scientific and engineering research and development. Some of its representative influences have occurred in biology, especially in the rise of ecology and genomics. Institutional changes in the social structure of knowledge, especially increased interdisciplinarity, have also resulted from systems thinking. Such boundary crossing has extended outside academia, as initiatives in systems-based problem solving opened up new lines of communication between scientists, engineers, and policy makers. Indeed, decision making can often be improved and unintended consequences minimized by integrating science and policy into systems where the supply and demand for different types of information is more tightly coupled than it might otherwise have been.

On the other hand, most of this work has limited itself to criticizing only the second or internalist level of disembedding. Moreover, it has undertaken this criticism in order to advance science and engineering in forms that remain on other grounds distinctly modern. While the means that characterize modern science once entailed methodological disembedding, such means were from the beginning developed in order to enhance human power. As the problems with which science has dealt became progressively more complex - that is, as the simple questions in physics and chemistry and biology answered - then science was forced to become less analytic or reductionist. Free body diagrams only work at a relatively gross level of analysis and construction, when limited methodological tools were available. As the problems became more complex and the tools (especially computers) more sophisticated, free body diagrams were forced to give way to more complex models - precisely in order to continue to advance the modern project. Insofar as systems thinking is proposed to enhance rather than delimit modern science and technology it has done so in order to enhance power.

It would thus be a mistake to separate systems science from its own political and cultural contexts. For example, Capra (1997) has argued that new research on the organization of living systems leads us to re-examine social policies. Systems thinking is both a scientific and a cultural paradigm shift away from mechanism and reductionism, but the relationship between these two shifts is complex and ethically charged. Capra's argument raises ethical questions about deriving political and moral conclusions (e.g., advocating for egalitarianism) from observations about nature. This is the same dilemma often raised by political conclusions drawn from the reductionist theories of sociobiology, which argue that ethics and values are themselves products of evolution (see, e.g., Wilson, 1998). The focus on wholeness, interconnectedness, and complexity has had an ambiguous impact on the larger landscape of cultural and philosophical thought. Although it has no doubt on occasion promoted a degree of scientific humility and criticism of technological power, it has also generated new versions of what Socrates, Plato, and Aristotle would have deemed theoretical and practical hubris.

3.   HUBRIS PRO AND CON

As a critique of technoscientific hubris, systems thinking interacted with changing conceptions of technology and the environment since the mid-20th century. This was evident, for instance, in the reception accorded Aldo Leopold's *A Sand County Almanac* (1949) and Rachel Carson's *Silent Spring* (1962). Although he did not use the term, Leopold essentially argued that the concept of system forms the foundation of ethics: "All ethics so far evolved rest upon a single premise: that the individual is a member of a community of interdependent parts" (p. 203). Systems thinking helped illuminate Leopold's ethical critique of technological hubris (that humans stand above and are able to dominate nature) by evincing and formalizing the many interconnections of human and natural systems. Carson explicitly rejects the "control of nature" as an attitude "conceived in arrogance" on the basis of a philosophical conception of nature as existing for human convenience (p. 297).

But systems thinking also led to a renewed hubris in the form of large-scale, technocratic interventions in human or natural systems. At the practical level, Robert McNamara explicitly credits his work in operations research and systems analysis as enhancing the U.S. bombing of civilian targets in Japan during World War II (Shapley, 1993); and during his tenure as the U.S. Secretary of Defense (1961-1968) this approach surely influenced decisions that escalated the Vietnam War. At the theoretical level, Luhmann's brand of systems thinking seeks to abstract a 'grand theory', or a universal framework that is not concerned about individual humans (only the abstractions of information exchange). This led Habermas (1981) to label it a version of 'anti-humanistic' sociology that denies the activity of individuals and institutions to consciously guide social change. Indeed, worldviews that stress holism always threaten to lose sight of individual values such as freedom, dignity, and intentionality. In this case modern societies are seen as polycentric, and democratic participation and control as illusory in the face of overwhelming complexity. Yet it is difficult to conceive of justice and many other social values materializing in the autopoietic process devoid of intentional agency as described by Maturana and Varela (1980).

Systems thinking has also led to critiques of scientific mechanism and reductionism. Adaptive management, based on flexible policies that are robust even in the face of persistent uncertainties, can be interpreted as forms of humility in the face of increased appreciation for the complexity of social-environmental systems. Yet the sophisticated computer models developed in part through systems theories can be used to reinforce a faith that it is possible to obtain the predictive knowledge necessary to manage the world as a human artifact, even while occasionally employing the rhetoric of moderation (Allenby, 2000). In the case of global climate change, the United States government has consistently delayed political action in favor of more research into climate systems so as to know better precisely what technocratic solutions to undertake.

Finally, not only has systems thinking exhibited an ambiguous heritage within science, engineering, and politics, but it has sometimes been used to promote radically new philosophical worldviews that are anything but devoid of hubris. For example, the web-based *Principia Cybernetica* project addresses perennial philosophical problems (e.g., the one and the many) with the principles derived from cybernetics and systems thinking. The holistic and integrative insights produced by systems thinking tie into Capra's notion (1975) that Western science is now challenging Cartesian dualism and the dominant view of a reality composed of fragmented, compartmentalized objects. Although some may draw spiritual fulfilment from such a vision of harmony between Western science and Eastern mysticism, the vision is also one that popularizes science. This raises interesting dilemmas about justifying greater scientific autonomy or funding in the name of spiritual enlightenment, not to mention the relationship between science and religion.

4.   THE HIERARCHY OF SYSTEM LEVELS

How does the Strijbos and Basden integrated vision for technology as located in a hierarchy of system levels fit into this typology? First it is clear, from the roots of this hierarchy in disclosive systems theory, that it aspires to be anti-hubris. Disclosive systems theory as developed by Strijbos draws on the work of the Dutch Calvinist philosopher Herman Dooyeweerd (1894-1977) to attempt a break with the dominant European idea of human autonomy as the basis for the pursuit of power over nature. Moreover, unlike other systems projects, this one clearly aims to deal with the issue of external as well as the internal disembedding. The humanities need to be reintegrated into technoscientific systems.

But in the hierarchy, technologies remain structurally as a kind of foundation, and only as one works up from this foundation through artifacts, human practices, and socio-technical systems does one come into contact with 'directional perspectives' that include religion, ethics, and the humanities. Why are ethics and the humanities, our understandings of and reflections on what it means to be human, not at the bottom, with technologies founded on them? Does this not reflect an acceptance, however subtle or unintentional, of our cultural assumption about the central and driving character of science and technology?

Consider a case study that complements others referenced in this volume. Constance Perin in *Shouldering Risks: The Culture of Control in the Nuclear Power Industry* (2005) examines four events at three nuclear power stations in the United States in order to analyze in detail the management of risks in these instances of what she calls 'heroic technologies'. Other examples are bullet trains, jumbo jets, manned space capsules, offshore oil rigs, and supertankers - all of which are "heroic in their ambitions, in the amazing materials making them possible, and in the immense amounts of money and the millions of people mobilized to design and operate them" (p. viii). Addressing the risks that are inevitably associated with such technologies requires maximizing communication in order to keep track of "expectable interactions within a complicated, often opaque system and responding promptly to those not expected" (p. xvi). Her key observation, in the words of one of the engineers, is that "We're technical people, but most of our problems are cultural" (p. xiii).

> "Maintaining the capacity for seeing in the round is a struggle of all specialists - technical, cultural, scientific. Refusing 'tunnel vision' and keeping their sights on a multidimensional world is especially difficult for those in these complicated enterprises doing their professional best in environments focused on costs and obsessed with time." (p. 281)

Obviously Perin's case studies provide limited confirmation of the hierarchies framework for understanding systems - and yet they do so in ways that can trivialize Strijbos and Basden's thesis. After all, what Perin is examining is a 'culture of control'. In such a culture, there is recognition of the importance of extending this culture beyond the technical. Moving from the inside of technologies out to the margins, scientists and engineers and managers are presented as recognizing the importance of taking into account not just artifacts but human practices, socio-technical systems, and even directional perspectives. But in all cases this is done with an eye toward enhancing power and control. Is there no need to question this fundamental culture? If one can make a case for so doing, would it not require making culture foundational rather than added? But in a culture in which technology is foundational and culture is additional, how could this possibly be done?

Consider again the radical criticism of the social ontology of systems found in Arney's *Experts in the Age of Systems* (1991). Going beyond Clarke (1961), who argued system as a new ontological category, Arney presents system as definitive of the social reality in which over the course of the 20th century the residents of advanced technological society have come progressively to exist. From his analysis of the lives of experts in the system that created the atomic bomb, Arney concludes that "complex systems are relentlessly resilient to change, rapidly responsive to criticism and remarkably stable for it" (pp. 4-5).

"Indeed, good, well-functioning systems depend on good criticism and use it well. Systems of expertise incorporate criticism in themselves. To try to be a critic is to double the expert and to become an agent of the system...." (p. 8)

The only adequate response, says Arney, is not to attempt some further refinement of systems thinking, but laughter.

Imagine that Mark Twain's *Connecticut Yankee* engineer were to reverse his time travel and wake up in one of our contemporary systems. Would he not discover himself more rather than less at home, and find it even more easy than in Arthurian England to move from his central commitment to technology to the critical transformation of a culture? In what ways would the Strijbos and Basden framework of hierarchies offer a stronger alternative to such incorporation than the laughter that Twain seeks to nourish in his readers? Such is at least one question to be addressed to their important proposal.

REFERENCES

Chapters are given in which each reference occurs.

Ackoff R.L. (1974). The social responsibility of Operational Research. *Operational Research Quarterly*, **25**:361-371. [ch.9]

Ackoff R.L. (1976). *Redesigning the Future: A Systems Appproach to Societal Problems*. John Wiley & Sons, New York / London. [ch.14]

Ackoff R.L. (1979a). The future of Operational Research is past. *Journal of the Operational Research Society*, **30**:93-104. [ch.13]

Ackoff R.L. (1979b). Resurrecting the future of Operational Research. *Journal of the Operational Research Society*, **30**:189-199. [ch.3]

Ackoff R.L. (1981). *Creating the Corporate Future: Plan or be Planned for*. John Wiley & Sons, New York / London. [ch.13,14]

Ackoff R.L. (1993). Idealized Design: creative corporate visioning. *International Journal of Management Science* **21**:401-410. [ch.3]

Ackoff R.L. (1994). *The Democratic Corporation: A Radical Prescription for Recreating Corporate America and Rediscovering Success*. Oxford University Press. [ch.12,14]

Allenby B.R. (2000). Earth Systems Engineering: the world as human artifact. *The Bridge* (U.S. National Academy of Engineering), **30**(1, Spring):5-13. [ch.16]

Arney W.R. (1991). *Experts in the Age of Systems*. University of New Mexico Press, Albuquerque, USA. [ch.16]

Ashby W.R. (1960). *Design for a Brain*, 2nd ed. Chapman & Hall, London. [ch.10]

Atthill C. (1975). *Decisions: West Oil Distribution*. PB Educational Service, London. [ch.13]

Attwood T. (2001). *Asperger's Syndrome: A Guide for Parents and Professionals*. Jessica Kingsley Publishers, London. [ch.4]

Augier M. and Vendelø M.T. (1999). Networks, cognition and management of tacit knowledge. *Journal of Knowledge Management*, **3**(4):252-261. [ch.4]

Avison D.E., Shah H.U., Golder P.A. (1993). Tools for SSM: a justification - a reply to 'Critique of two contributions to soft systems methodology'. *European Journal of Information Systems* **2**:312-313. [ch.5]

Bahm N.K. (1995). The emergence of community in computer-mediated communication. *Cybersociety - Computer-mediated Communication and Community*, S.G. Jones, ed., Sage Publications, London / New Delhi, pp. 138-163. [ch.8]

Bailey K.D. (2001). Towards unifying science: Applying concepts across disciplinary boundaries. *Systems Research and Behavioral Science*, **18**:41-62. [ch.15]

Basden A. (1993). Appropriateness. *Research and Development in Expert Systems X; Proc. Expert Systems 93*, M.A. Bramer, A.L. Macintosh eds., BHR Group, UK, pp.315-328. [ch.2, Preface]

Basden A. (1994). Three levels of benefit in expert systems. *Expert Systems*, **11**(2):99-107. [ch.11, Preface]

Basden A. (2001). A philosophical underpinning for I.T. evaluation. *8th European Conference on IT Evaluation, ECITE*. Oriel College, Oxford, UK, 17-18 September 2001. [ch.11, Preface]

Basden A. (2002). A philosophical underpinning for I.S. development. *Proceedings of the 10th European Conference on Information Systems (ECIS)* Gdansk, Poland. [ch.5]

Basden A. and Brown A.J. (1996). Istar - a tool for creative design of knowledge bases. *Expert Systems*, **13**(4):259-276. [ch.2]

Basden A. and Hibberd P.R. (1996). User interface issues raised by knowledge refinement. *Int. J. Human Computer Studies*, **45**:135-155. [ch.2]

Basden A. and Wood-Harper A.T. (2005). A philosophical discussion of the Root Definition in Soft Systems Thinking: An enrichment of CATWOE. *System Research and Behavioral Science*, **22**:1-27. [ch.11]

Baskerville R.L. and Wood-Harper A.T. (1996). A critical perspective on action research as a method for information systems research. *Journal of Information Technology*, **11**:235-246. [ch.3]

Baskerville R.L. and Wood-Harper A.T. (1998). Diversity in information systems action research methods. *European Journal of Information Systems*, **7**:90-107. [ch.3]

Bateson G. (1972). *Steps to an Ecology of Mind*. Jason Aronson, Northvale NJ, USA. [ch.15]

Bauman Z. (1993). *Postmodern Ethics*. Blackwell Publishers, Oxford, UK. [ch.7]

Bausch K.C. (1997). The Habermas/Luhmann debate and subsequent Habermasian perspectives on Systems Theory. *Systems Research and Behavioral Science*, **14**:315-330. [ch.14]

Beck U. (1992). *Risk Society - Towards a New Modernity*. Sage Publications, London / New Delhi. [ch.7,8]

Beer S. (1979). *The Heart of Enterprise*. Wiley, New York. [ch.10]

Beer S. (1981). *Brain of the Firm*, 2nd ed. Wiley, New York. [ch.10]

Beer S. (1959). *Cybernetics and Management*. Wiley, New York. [ch.16]

Bell D. (1973). *The Coming of the Post-Industrial Society: A Venture in Social Forecasting*. Basic Books, New York. [ch.6]

Bell D. (1979). The social framework of the Information Society. *The Computer Age: A twenty Year View*, M. Dertouzos, J. Moses, eds., MIT Press, Cambridge, MA, USA, pp. 163-212. [ch.6]

Berger P.L. (1979). *The Heretical Imperative - Contemporary Possibilities of Religious Affirmation*. Anchor Press, Doubleday, Garden City, New York. [ch.7]

Bergvall-Kåreborn B. (2000). *Using Soft Systems Methodology as a Methodology for Multi-Modal Systems Design*, Licentiate Thesis. Luleå University of Technology, Sweden. [ch.10]

Bergvall-Kåreborn B. (2001). The role of the Qualifying Function concept in systems design. *Systemic Practice and Action Research*, **14**:79-93. [ch.3, Preface]

Bergvall-Kåreborn B. (2002a). Qualifying Function in SSM modelling - a case study. *Systemic Practice and Action Research*, **15**:309-330. [ch.3, Preface]

Bergvall-Kåreborn B. (2002b). *A Multi-Modal Approach to Soft Systems Methodology*, Doctoral Thesis. Luleå University of Technology, Sweden. [ch.5]

Bergvall-Kåreborn B. (2002c). Enriching the model-building phase of Soft Systems Methodology. *Systems Research and Behavioral Science*, **19**:27-48. [Preface]

Bergvall-Kåreborn B. and Grahn A. (1996a). Multi-Modal Thinking in Soft Systems Methodology's rich pictures. *World Futures*, **47**:79-92. [ch.5,10, Preface]

Bergvall-Kåreborn B. and Grahn A. (1996b). Expanding the framework for monitor and control in Soft Systems Methodology. *Systems Practice*, **9**(5):469-495. [ch.3,5,10, Preface]

Bergvall-Kåreborn B., Mirijamdotter A., and Basden A. (2004). Basic principles of SSM modeling: an examination of CATWOE from a soft perspective. *Systemic Practice and Action Research*, **17**(2):55-73. [ch.5]

Berners-Lee T,. Fielding R., Frystyk H. (1996). *RFC 1945 Hypertext Transfer Protocol -*

HTTP/1.0. Internet Engineering Task Force, "ftp://ftp.isi.edu/in-notes/rfc1945.txt". [ch.1]

Blaha M., Premerlani W. (1998). *Object-Oriented Modelling and Design for Database Applications.* Prentice-Hall, NJ, USA. [ch.2]

Boden M.A. (1999). What is interdisciplinarity? *Interdisciplinarity and the Organisation of Knowledge in Europe*, R. Cunningham, ed., Office for Official Publications of the European Communities, Luxembourg, pp. 13-24. [ch.1,4,15]

Bogdanov A.A. (1913-1917). *Bogdanov's Tektology*, 1996 ed. Dudley P. Centre for Systems Studies Press, Hull, UK. [ch.15]

Borgmann A. (1999). *Holding on to Reality - The Nature of Information at the Turn of the Millennium.* The University of Chicago Press, Chicago, London. [ch.8]

Boulding K.E. (1953). *The Organizational Revolution: A Study in the Ethics of Economic Organization.* Harper & Row, New York. [ch.6]

Boulding K.E. (1956). General systems theory - the skeleton of science. *Management Science*, **2**(3):197-208. [ch.10,15]

Boulding K.E. (1956). *The Image: Knowledge in Life and Society.* The University of Michigan Press, Michigan, USA. [ch.3]

Bourdieu P. (1977). *Algérie 60 - Structures économiques et Structures Temporelles.* Les Éditions de Minuit, Paris. [ch.7]

Bourdieu P. (1980). *Le Sens Pratique.* Les Éditions de Minuit, Paris. [ch.7]

Bourdieu P. (1986). L'illusion biographique. *Actes de la Recherche en Sciences Sociales*, **62/63**:69-72. [ch.7]

Bourdieu P. (1991). *Language and Symbolic Power.* Polity Press, Cambridge, UK. [ch.7,8]

Bourdieu P. (2000). *Pascalian Meditations.* Polity Press, Cambridge, UK. [ch.7]

Bowen K.C. (1986). An eighth face of research. *Omega*, **18**(2). [ch.13]

Brachman R.J. (1990). The future of knowledge representation. *AAAI-90, Proc. Eighth National Conference on Artificial Intelligence*, pp. 1082-92. [ch.2]

Brandon P.S., Basden A., Hamilton I., Stockley J. (1988). *Expert Systems: Strategic Planning of Construction Projects.* The Royal Institution of Chartered Surveyors, London, UK. [ch.11]

Brandon P.S. and Lombardi P. (2005). *Evaluating Sustainable Development in the Built Environment.* Blackwell Science, Oxford, UK . [ch.11]

Bunge M. (1977). *Treatise on Basic Philosophy, Volume 3: Ontology 1: The Furniture of the World.* Reidal, Boston, USA. [ch.2]

Bunge M. (1979). *Treatise on Basic Philosophy, Volume 4: Ontology 2: A World of Systems.* Reidal, Boston, USA. [ch.2,11]

Burill J. (1987). *Kritisk Teori - en Introduktion*, [Critical Theory - An Introduction]. Diablos, Lund, Sweden. [ch.10]

Burrell G. and Morgan G. (1979). *Sociological Paradigms and Organizational Analysis.* Heinemann, London. [ch.13]

Butterfield J., Pendegraft N. (1996). Cultural analysis in IS planning & management. *Journal of Systems Management*, **47**:14-17. [ch.1,11]

Callo V.N., Packham R.G. (1999). The use of Soft Systems Methodology in emancipatory development. *Systems Research and Behavioral Science*, **16**:311-319. [ch.3]

Capra F. (1975). *The Tao of Physics: An Exploration of the Parallels Between Modern Physics and Eastern Mysticism.* Bantam Books, New York. [ch.16]

Capra F. (1997). *The Web of Life: A New Scientific Understanding of Living Systems.* Anchor Books, New York. [ch.16]

Carson R. (1962). *Silent Spring*. Houghton Mifflin, Boston. [ch.16]

Castell A.C., Basden A., Erdos G., Barrows P., Brandon P.S. (1992). Knowledge based systems in use: a case study. *Proc. Expert Systems 92 (Applications Stream)*. British Computer Society Specialist Group for Knowledge Based Systems. UK. [ch.11]

Castells M. (1996). *The Information Age: Economy, Society and Culture, Vol. I: The Rise of the Network Society*. Blackwell Publishers, Oxford, UK. [ch.6]

Castells M. (2002). *The Internet Galaxy - Reflections on the Internet, Business, and Society*. Oxford University Press. [ch.7]

Checkland P.B. (1970). Systems and science, industry and innovation. *Journal of Systems Engineering*, 1:29-44. [ch.1]

Checkland P.B. (1971). A systems map of the universe. *Journal of Systems Engineering*, 2:107-114. [ch.3]

Checkland P.B. (1972). Towards a systems-based methodology for real-world problem solving. *Journal of Systems Engineering*, 3:87-116. [ch.5]

Checkland P.B. (1978). The origins and nature of 'hard' systems thinking. *Journal of Applied Systems Analysis*, 5:99-110. [ch.1,9,13,14]

Checkland P.B. (1979a). The problem of problem formulation in the application of a systems approach. *Education in Systems Science*, B.A. Bayraktar, H. Müller-Merbach, J.E. Roberts, M.G. Simpson, eds., Taylor and Francis, London, pp. 318-326. [ch.3]

Checkland P.B. (1979b). Techniques in 'soft' systems practice; Part 2: building conceptual models. *Journal of Applied Systems Analysis*, 6:41-49. [ch.3,5]

Checkland P.B. (1981). *Systems Thinking, Systems Practice*. Wiley, Chichester, UK. [ch.1,3,4,5,9,10,11,13]

Checkland P.B. (1982). Soft Systems Methodology as process: a reply to M. C. Jackson. *Journal of Applied Systems Analysis*, 9:37-39. [ch.3]

Checkland P.B. (1984). Systems theory and information systems. *Beyond Productivity: Information Systems Development for Organizational Effectiveness*, T.M.A. Bemelmans, ed., Elsevier Science Publishers B V., North-Holland, Amsterdam, pp. 9-21. [ch.3]

Checkland P.B. (1985). From optimizing to learning: a development of systems thinking for the 1990s. *Journal of the Operational Research Society*, 36:757-767. [ch.3,9,13]

Checkland P.B. (1988). Information systems and systems thinking: time to unite? *International Journal of Information Management*, 8:239-248. [ch.3]

Checkland P.B. (1989). Soft Systems Methodology. *Rational Analysis for a Problematic World*, J. Rosenhead, ed., Wiley, Chichester, UK, pp. 71-100. [ch.5]

Checkland P.B. (1991). From optimizing to learning: a development of systems thinking for the 1990s. *Critical Systems Thinking*, R.L. Flood, M.C. Jackson, eds., John Wiley & Sons, Chichester, UK. [ch.3]

Checkland P.B. (1995). Model validation in soft systems practice. *Systems Research*, 12:47-54. [ch.3]

Checkland P.B. (1999). *Systems Thinking, Systems Practice: Includes a 30-Year Retrospective*. Wiley, Chichester, UK.. [ch.5,11]

Checkland P.B. (2000a). New maps of knowledge. *Systems Research and Behavioral Science*, 17:59-77. [ch.1]

Checkland P.B. (2000b). Soft Systems Methodology: a thirty year retrospective. *Systems Research and Behavioral Science*, 17(S1):S11-S58. [ch.5]

Checkland P.B. and Davies L. (1986). The use of the term 'Weltanschauung' in Soft Systems Methodology. *Journal of Applied Systems Analysis*, 13:109-115. [ch.3]

Checkland P.B., Forbes, P., Martin, S. (1990). Techniques in soft systems practice; Part 3:

monitoring and control in conceptual models and in evaluation studies. *Journal of Applied Systems Analysis*, **17**:29-37. [ch.5]

Checkland P.B. and Griffin R. (1970). Management information systems: a systems view. *Journal of Systems Engineering*, **1**:29-43. [ch.3]

Checkland P.B. and Holwell S. (1993). Information management and organizational processes: an approach through Soft Systems Methodology. *Journal of Information Systems*, **3**:3-16. [ch.3,5]

Checkland P.B. and Holwell S. (1997). *Information, Systems and Information Systems*. John Wiley & Sons, Chichester, UK. [ch.3]

Checkland P.B. and Holwell S. (1998). Action research: its nature and validity. *Systemic Practice and Action Research*, **11**:9-21. [ch.3]

Checkland P.B. and Scholes J. (1990a). *Soft Systems Methodology in Action*. John Wiley & Sons, New York. [ch.3,5,10]

Checkland P.B. and Scholes J. (1990b). Techniques in soft systems practice part 4: conceptual model building revisited. *Journal of Applied Systems Analysis*, **17**:39-43. [ch.3]

Checkland P.B. and Scholes J. (1999). *Soft Systems Methodology in Action: Includes a 30-Year Retrospective*. John Wiley & Sons, Chichester, UK. [ch.3,5]

Checkland P.B. and Tsouvalis, C. (1997). Reflecting on SSM: the link between root definitions and conceptual models. *Systems Research and Behavioral Science*, **14**:153-168. [ch.3]

Checkland P.B. and Wilson, B. (1980). 'Primary task' and 'issue-based' root definitions in systems studies. *Journal of Applied Systems Analysis*, **7**:51-54. [ch.3]

Churchman C.W. (1948). *Theory of Experimental Inference*. Macmillian, New York. [ch.13]

Churchman C.W. (1968a). *Challenge to Reason*. McGraw-Hill, New York. [ch.13]

Churchman C.W. (1968b,1984). *The Systems Approach*. 1968: Delacorte Press, New York, 1984: Dell Publishing Co., New York. [ch.3,9,13,16]

Churchman C.W. (1970). Operations Research as a profession. *Management Science*, **17**(2):37-53. [ch.13]

Churchman C.W. (1971). *The Design of Inquiring Systems: Basic Concepts of Systems and Organization*. Basic Books, London. [ch.3,13]

Churchman C.W. (1979). *The Systems Approach and Its Enemies*. Basic Books, New York. [ch.13]

Churchman C.W. (1982). *Thought and Wisdom*. Intersystems Publications, Seaside, CA, USA. [ch.13]

Churchman C.W. (1987). Systems profile: discoveries in an exploration into systems thinking. *Systems Research*, **4**(22):139-146. [ch.13]

Churchman C.W., Ackoff R.L., Arnoff E.L. (1957). *Introduction to Operations Research*. Wiley, New York. [ch.13]

Clarke W.N. (1961). System: a new category of Being. *Proceedings of the Twenty-third Annual Convention of the Jesuit Philosophical Association*, Woodstock College Press, Woodstock, NY, USA, pp. 5-17. [ch.16]

Collins R. (1975). *Conflict Sociology: Toward an Explanatory Approach*. Academic Press, New York. [ch.12]

Coolen M. (1997). Totaal verknoopt - Internet als verwerkelijking van het moderne mensbeeld. *Virtueel verbonden - Filosoferen over cyberspace*, V. De Boer, J. Vorstenbosch, eds., Parrèssia, Amsterdam, pp. 28-59. [ch.8]

Cotterill T. and Law N. (1993). EIS - a practical approach. *HPECU/HPCUA Proceedings of the 1993 Hewlett-Packard Computer Users' European Conference.* HP Computer Users Assoc. [ch.1,11]

Córdoba J.R. and Midgley G. (2003). Addressing organisational and societal concerns: An application of critical systems thinking to information systems planning in Colombia. *Critical Reflections on Information Systems: A Systemic Approach,* J.J. Cano ed., Idea Group, Hershey, UK. [ch.15]

Dahlbom B. (1992). The idea that reality is socially constructed. *Software Development and Reality Construction,* C. Floyd, H. Züllighoven, eds., Springer-Verlag, Berlin, pp. 101-126. [ch.3]

Dahlbom B. (2000). Nätverkande: om organisering och ledning i e-samhället. *IT, Organiserande och Ledarskap,* K. Ydén, ed., Bokförlaget BAS, Göteborg, Sweden, pp. 111-138. [ch.5]

De Raadt J.D.R. (1989). Multi-Modal Systems Design: a concern for the issues that matter. *Systems Research,* 6(1):17-25 [ch.10]

De Raadt J.D.R. (1991). *Information and Managerial Wisdom.* Paradigm Publications, Idaho, USA. [ch.5,10,13]

De Raadt J.D.R. (1995). Expanding the horizon of information systems design: information technology and cultural ecology. *Systems Research,* 12:185-199. [ch.5,9,13]

De Raadt J.D.R. (1996). What the prophet and the philosopher told their nations: a multi-modal systems view of norms and civilisation. *World Futures,* 47:53-67. [ch.13]

De Raadt J.D.R. (1997a). Faith and the normative foundation of systems science. *Systems Practice,* 10(1):13-35. [ch.13]

De Raadt J.D.R. (1997b). A sketch for humane operational research in a technological society. *Systems Practice,* 10(4):421-41. [ch.11]

De Raadt J.D.R. (2000). *Redesign and Management of Communities in Crisis.* Universal/uPublish.com, USA. [ch.10]

De Raadt J.R.D. (1998). A new management of life. *Toronto Studies in Theology,* 75. The Edwin Mellen Press, New York. [ch.10]

De Smet P.A.G.M. (1988). The Dutch approach to computerized drug information: conceptual basis and realization. *J. Soc. Adm. Pharm.,* 5:49-58. [ch.9]

Descartes R. (1983). *Discours de la Méthode.* Éditions sociales, Paris. [ch.7]

Dooyeweerd H. (1953,1955,1958,1969,1975,1984,1997). *A New Critique of Theoretical Thought, Volumes I-IV.* 1953, 1955, 1958, 1969: The Presbyterian and Reformed Publishing Company, Philadelphia, USA.; 1975, 1984: Paideia Press (1975 edition), Jordan Station, Ontario, Canada, 1997: The Edwin Mellen Press, Lewiston, USA. [ch.1,2,3,4,5,9,10,11,12,13,14,15]

Dooyeweerd H. (1968). *In the Twilight of Western Thought: Studies in the Pretended Autonomy of Philosophical Thought.* The Craig Press, Nutley, New Jersey, USA. [ch.14]

Dooyeweerd H. (1973). Introduction. *Philosophia Reformata,* 38:5-16. [ch.3]

Dooyeweerd H. (1979). *Roots of Western Culture: Pagan, Secular and Christian Options.* Wedge Publishing Foundation, Toronto. [ch.14]

Duisterhout J.S., Van der Meulen A.F., Boersma J.J., Gebel R.S., Njoo K.H. (1992). Implementation of ICPC coding in information systems for primary care. *Medinfo,* 92:1483-1488. [ch.9]

Earl M. (2001). Knowledge management strategies: towards a taxonomy. *Journal of Management Information Systems,* 18(1):215-233. [ch.4]

Eden C. (1988). Cognitive Mapping. *European Journal of Operational Research,* 36:1-13. [ch.4]

Eden C. and Radford J. (1990). *Tackling Strategic Problems*. Sage, London. [ch.13]

Einstein A. (1934). *The world as I see it*. Philosophical Library, New York 1949. [Translation of *Mein Weltbild*. Frankfurt am Main, Ullstein 1934.] [ch.9]

Ellul J. (1964). *The Technological Society*. Alfred A. Knopf, New York. [ch.6,12]

Ellul J. (1980a). *The Technological System*. The Continuum Publishing. Corporation, New York. [ch.6,12]

Ellul J. (1980b). Nature, technique and artificiality. *Research in Philosophy and Technology*, P.T. Durbin ed., JAI Press Inc, Greenwich, CT, USA. [ch.12]

Ellul J. (1980c). The power of technique and the ethics of non-power. *The Myths of Information: Technology and Post-Industrial Culture*, K. Woodward ed., Routledge and Kegan Paul, London. [ch.12]

Ellul J. (1986). *The Ethics of Freedom*. Eerdmans, Grand Rapids, MI, USA. [ch.12]

Eriksson D.M. (1998). *Managing Problems of Postmodernity: Some Heuristics for Evaluation of Systems Approaches*. Interim report IR-98-060/August. International Institute for Applied Systems Analysis. A-2361 Laxenburg, Austria. [ch.10,13]

Eriksson D.M. (2001). Multi-modal investigation of a business process and information system redesign: a post-Implemetation case study. *Systems Research and Behavioral Science*, 18(2):181-196. [ch.5, Preface]

Eriksson D.M. (2003). An identification of normative sources for systems thinking: an inquiry into religious ground-motives for systems thinking paradigms. *Systems Research & Behavioral Science*, 20(6):475-487. [Preface]

ESRC Global Environmental Change Programme (1999). *The Politics of GM Food: Risk, Science and Public Trust. Special Briefing #5*. University of Sussex, Brighton, UK. [ch.15]

Fahey L. and Prusak L. (1998). The eleven deadliest sins in knowledge management. *California Management Review*. Spring 1998. [ch.4]

Fairtlough G. (1982). A note on the use of the term 'Weltanschauung' (W) in Checkland's Systems Thinking, Systems Practice. *Journal of Applied Systems Analysis*, 9:131-132. [ch.3]

Finin T., Labrou Y., Mayfield J. (1997). KQML as an agent communication language. *Software Agents*, J. Bradshaw ed., MIT Press, Cambridge, MA, USA. [ch.1]

Flood R.L. and Carson E.R. (1993). *Dealing with Complexity. An Introduction to the Theory and Application of Systems Science*, 2nd ed. Plenum, New York. [ch.10]

Flood R.L. and Jackson M.C. (1991a). *Creative Problem Solving - Total Systems Intervention*. John Wiley & Sons, Chichester, UK. [ch.3,10,13]

Flood R.L. and Jackson M.C. (1991b). *Critical Systems Thinking - Directed Readings*. John Wiley & Sons, Chichester, UK. [ch.3,10,13]

Flood R.L. and Romm, N.R.A. (1996a). *Critical Systems Thinking. Current Research and Practice*. Plenum, New York, USA. [ch.10]

Flood R.L. and Romm, N.R.A. (1996b). *Diversity Management. Triple Loop Learning*. Wiley, Chichester, UK. [ch.10,13]

Flood R.L., and Ulrich, W. (1990). Testament to conversations on Critical Systems Thinking between two systems practitioners. *Systems Practice*, 3:7-29. [ch.3]

Forrester J.W. (1961). *Industrial Dynamics*. Wright-Allen Press, Cambridge. [ch.13]

Franklin U. (1990). *The Real World of Technology*. CBC Enterprises, Toronto. [ch.6]

Friedman T. (1995). Making sense of software - computer based games and interactive textuality. *Cybersociety - Computer-mediated Communication and Community*, S.G. Jones, ed., Sage Publications, London, New Delhi, pp. 73-89. [ch.8]

Funt B.V. (1980). Problem-solving with diagrammatic representations. *Artificial Intelligence*, **13**(3):201-230. [ch.2]

Galliers R.D., and Land F.F. (1987). Choosing appropriate information systems research methodologies. *Communcations of the ACM*, **30**:900-902. [ch.3]

Gass S.I., Harris, C.M. (1996). *Encyclopedia of Operations Research and Management Science*. Kluwer Academic Publications. Boston. [ch.13]

Geertsema H.G. (1992). *Het Menselijk Karakter van ons Kennen* [The Human Character of our Knowing]. Buijten & Schipperheijn, Amsterdam. [ch.14]

Geertsema H.G. (1993). Homo respondens: on the historical nature of human reason. *Philosophia Reformata*, **58**:120-52. [ch.14]

Genesereth M.R., Fikes R. (1992). *Knowledge Interchange Format, Version 3.0 Reference Manual*, Technical Report Logic-92-1, June 1992. Computer Science Department, Stanford University, USA. [ch.1]

Giddens A. (1990). *The Consequences of Modernity*. Polity Press, Cambridge. [ch.7]

Giddens A. (1991). *Modernity and Self-Identity - Self and Society in the Late Modern Age*. Stanford University Press, Stanford, CA, USA. [ch.7,8]

Gladden G.R. (1982). Stop the life-Cycle, I want to get off. *ACM SIGSOFT Software Engineering Notes*, **7**(2):35-39. [ch.1,11]

Glas G. (1995). Ego, self, and the body. an assessment of Dooyeweerd's philosophical anthropology. *Christian Philosophy at the Close of the Twentieth Century: Assessment and Perspective*, S. Griffioen, B.M. Balk, eds., Uitgeverij Kok, Kampen, Netherlands, pp. 67-78. [ch.3]

Glaser B.G. and Strauss A.L. (1967). *The Discovery of Grounded Theory: Strategies for Qualitative Research*. Aldine de Gruyter, New York. [ch.4]

Goudzwaard B. (1997). *Capitalism and Progress: A Diagnosis of Western Society*. Paternoster Press / William B. Eerdmans, Grand Rapids, MI, USA. [ch.14]

Gouldner A.W. (1975). *The Dark Side of the Dialectic: Toward a New Objectivity*. Sociology Institute of Amsterdam, Amsterdam. [ch.15]

Graham G. (1999). *The Internet - A Philosophical Inquiry*. Routledge, London, New York. [ch.8]

Green-Armytage J. (1993). Why Taurus was always ill-starred. *Computer Weekly*, 18 March, 10. [ch.1]

Greeno J. (1994). Gibson's affordances. *Psychological Review*, **101**:336-342. [ch.2]

Gregory W.J. (1992). *Critical Systems Thinking and Pluralism: A New Constellation*. Ph.D. thesis, City University, London. [ch.15]

Groth L. (1999). *Future Organizational Design - The Scope for the IT-based Enterprise*. John Wiley & Sons, Chichester, UK. [ch.3]

Habermas J. (1972). *Knowledge and Human Interests*, tr. Shapiro J.J. Heinemann, London. [ch.4]

Habermas J. (1981). *Theorie des Kommunikativen Handelns, vol. 2: Zur Kritik der Funktionalistischen Vernunft*. Suhrkamp, Frankfurt am Main. English version of the 3rd corrected edition, 1989: *The Theory of Communicative Action, vol. 2: Lifeworld and System: A Critique of Functionalist Reason*, tr. McCarthy T. Beacon Press, Boston, USA. [ch.16]

Habermas J. (1983). *Moralbewustsein und Kommunikatives Handeln*. Suhrkamp Verlag, Frankfurt am Main. [ch.8]

Habermas J. (1984). *The Theory of Communicative Action - Volume One: Reason and the Rationalization of Society*, tr. McCarthy T., Polity Press. [ch.4,10,13,15]

Habermas J. (1987). *The Theory of Communicative Action - Volume Two: The Critique of Functionalist Reason*. Polity Press, Cambridge. [ch.10,13,15,(16)]

Habermas J. (1994). *Faktizität und Geltung - Beiträge zur Diskurstheorie des Rechts und des Demokratischen Rechtsstaats*. Suhrkamp Verlag, Frankfurt am Main (vierte erweiterte Auflage). [ch.7,8]

Ham M. ten (1992). WHO's role in international ADR monitoring. *Post Marketing Surveillance*, **5**:223-230. [ch.9]

Hart H. (1984). *Understanding our World - An Integral Ontology*. University Press of America, Lanham, New York / London. [ch.8,11]

Health Council (1993). *Privacy and Postmarketing Surveillance* (in Dutch). Health Council, Gravenhage, The Netherlands. [ch.9]

Heim M. (1993). *The Metaphysics of Virtual Reality*. Oxford University Press, New York, Oxford. [ch.8]

Henderson R.D. (1994). *Illuminating Law; The Construction of Herman Dooyeweerd's Philosophy*. Free University, Amsterdam. [ch.2]

Hibberd P. and Basden A. (1995). Procurement and the use of intelligent systems of contract authoring. *Proc. RICS COBRA Conference*, September 1995, Edinburgh. [ch.2]

Hirschheim R. and Klein H. (1989). Four paradigms of information systems development. *Communications of the ACM*, **32**:1199-1216. [ch.10]

Hirschheim R. and Klein H. (1994). Realizing emancipatory principles in information systems development: The case for ETHICS. *MIS Quarterly*, **18**:83-109. [ch.10]

Hirschheim R., Klein H.K., Lyytinen K. (1996). Exploring the intellectual structures of information systems development: A social action theoretic analysis. *Accounting, Management & Information Technology*, **6**(1-2):1-64. [ch.10]

Holst M. (2003). Knowledge management and the concept of Ba: designing places for interaction across traditional organizational boundaries. *Proceedings of 26th Information Systems Research Seminars in Scandinavia, IRIS 26*, Haikko, Finland. [ch.5]

Holst M. (2004). *Knowledge Work across Boundaries - Inquiring into the Processes of Creating a Shared Context*. Licentiate Thesis, Department of Business Administration and Social Sciences, Luleå University of Technology, Sweden. http://epubl.luth.se/1402-1757/2004/66/index.html [ch.5]

Holst M. and Mirijamdotter A. (2004a). Logically created organizations for multidisciplinary settings: understanding their systemic structure. *Proceedings of the Third International Conference on Systems Thinking in Management*. Philadelphia, USA, May 19-21, 2004. http://www.acasa.upenn.edu/icstm04/ [ch.5]

Holst M. and Mirijamdotter A. (2004b). The creation of a shared context in a multidisciplinary setting. *Proceedings of The Fourth International Conference on Knowledge, Culture and Change in Organisations*. University of Greenwich, London. [ch.5]

Holst M., Mirijamdotter A., Bergvall-Kåreborn B., Oskarsson H. (2004). Information and communication technology in dynamic organisations. *Proceedings of 27th Information Systems Research Seminars in Scandinavia, IRIS 27*. Falkenberg, Sweden. http://w3.msi.vxu.se/users/per/IRIS27/iris27-1046.pdf [ch.5]

Hughes T.P. (1989). *American Genesis: A Century of Invention and Technological Enthusiasm, 1870-1970*. Viking, New York. [ch.16]

Hughes T.P. (2004). *Human-Built World: How to Think about Technology and Culture*. University of Chicago Press, Chicago. [ch.16]

Hult M. and Lennung S. (1980). Towards a definition of action research: a note and bibliography. *Journal of Managment Studies*, **17**:241-250. [ch.3]

Ihde D. (2002). *Bodies in Technology*. University of Minnesota Press, Minneapolis, USA. [ch.7]

Iivari J., Hirschheim R., Klein H.K. (1998). A paradigmatic analysis contrasting information systems development approaches and methodologies. *Information Systems Research*, 9(2):164 -193. [ch.10]

Ivanov K. (1991). Critical Systems Thinking and information technology: some summary reflections, doubts, and hopes through Critical Thinking critically considered, and through hypersystems. *Journal of Applied Systems Analysis*, 18:39-55. [ch.3]

Ivanov K. (1996). Presuppositions in information systems design: from systems to networks and contexts. *Accounting, Management & Information Technology*, 6(1/2):99-113. [ch.10]

Jackson M.C. (1982). The nature of 'soft' systems thinking: the work of Churchman, Ackoff and Checkland. *Journal of Applied Systems Analysis*, 9:17-29. [ch.3,13,14]

Jackson M.C. (1985). Social systems theory and practice: the need for a critical approach. *International Journal of General Systems*, 10:135-151. Gordon & Breach Science Publishers Inc, London. [ch.9]

Jackson M.C. (1991). *Systems Methodology for the Management Sciences*. Plenum Press, New York, London. [ch.4,9,13,14,15]

Jacob M. and Ebrahimpur G. (2001). Experience vs. expertise: the role of implicit understanding of knowledge in determining the nature of knowledge transfer in two companies. *Journal of Intellectual Capital*, 2(1):74-88. [ch.4]

Jayaratna N. (1994). *Understanding and Evaluating Methodologies - NIMSAD A Systemic Framework*. McGraw-Hill Book Company, London. [ch.5]

Jenkins G.M. (1969). The systems approach. *Journal of Systems Engineering*, 1(1):3-49. [ch.9]

Jenkins, L. (1969). The systems approach. *Systems Behaviour*, J. Beishon, G. Peters eds., 2nd ed, Harper & Row, New York. [ch.13]

Johannessen, J., Olaisen, J., Olsen, B. (2001). Mismanagement of tacit knowledge: the importance of tacit knowledge , the danger of information technology, and what to do about it. *International Journal of Information Management*, 21(2001):3-20. [ch.4]

Jonas H. (1974). *Philosophical Essays: From Ancient Creed to Technological Man*. Prentice-Hall, Englewood Cliffs, NJ, USA. [ch.12]

Jonas H. (1979). *Das Prinzip Verantwortung: Versuch einer Ethik für die Technologische Zivilisation*. Insel Verlag, Frankfurt am Main. [ch.12]

Kakabadse N.K., Kouzmin, A., Kakabadse, A. (2001). From tacit knowledge to knowledge management: leveraging invisible assets. *Knowledge and Process Management*, 8(3):137-154. [ch.4]

Kalsbeek L. (1975). *Contours of a Christian Philosophy - An Introduction to Herman Dooyeweerd's Thought*. Wedge Publishing Foundation, Toronto. [ch.3,5,11]

Kant I. (1933 [1787]). *The Critique of Pure Reason*, 2nd ed, tr. N.K. Smith Macmillan, Basingstoke, UK. [ch.15]

Kant I. (1987). *Critique of Judgement*. tr. Pluhar W.S. Hackett, Indianapolis, USA. [ch.13]

Kant I. (1997). *Critique of Practical Reason*, tr. Gregor M. Cambridge University Press. [ch.13]

Kant I. (1998). *Critique of Pure Reason*. tr. P. Guyer, A.W. Wood. Cambridge: Cambridge Univ. Press. [ch.13]

Kapp E. (1978 [1877]). *Grundlinien einer Philosophie der Technik: Zur Entstehungsgeschichte der Cultur aus neuen Gesichtspunkten*. Stern-Verlag Janssen & Co, Düsseldorf, Germany. [ch.6]

Kartowisastro H. and Kijima K. (1994). An enriched soft systems methodology and its application to cultural conflict under a paternalistic value system. *Systems Practice*, 7:241-253. [ch.3]

Kendall J. and Avison D. (1993). Emancipatory research themes in information systems development: human, organizational and social aspects. *Human, organizational and social dimensions of information systems development*, D. Avison, J. Kendall, J. DeGross, eds., North-Holland, Amsterdam, pp. 1-12. [ch.10]

Kierkegaard S. (1978). *Two Ages - The Age of Revolution and the Present Age - A Literary Review*. Princeton University Press; Princeton, New Jersey, USA. [ch.7]

Klein H.K. and Hirscheim, R. (1993). The application of neohumanist principlesin information systems development. *Human, organizational and social dimensions of information systems development*, D. Avison, J. Kendall, J. DeGross, eds., North-Holland, Amsterdam, pp. 263-280. [ch.10]

Kroesen J.O. (1999). Book review of Hans Haaksma et al., 'Van Riessen, filosoof van de techniek'. Leende, 1999. *Philosophia Reformata*, **64**(2):165-167. [ch.6]

Kuhn T.S. (1970). *The Structure of Scientific Revolutions*, University of Chicago Press. [ch.13]

Köhler V. and Bergvall-Kåreborn, B. (2004). Artefacts and actions - organizational transformation through use of IT? *Proceedings of the Third International Conference on Systems Thinking in Management*. Philadelphia, USA, May 19-21, 2004. http://www.acasa.upenn.edu/icstm04/ [ch.3]

Lamberts H. and Wood M. (1987). *International Classification of Primary Care*. Oxford University Press. [ch.9]

Landauer T.K. (1996). *The Trouble with Computers: Usefulness, Usability and Productivity*. Bradford Books, MIT Press, Cambridge, MA, USA. [ch.1]

Lang J.C. (2001). Managerial concerns in knowledge management. *Journal of Knowledge Management*, 5(1):43-57. [ch.4]

Lash S. and Urry J. (1994). *Economies of Signs and Spaces*. Sage Publications, London. [ch.8]

Laszlo E. (1972a). *The Systems View of the World: The Natural Philosophy of the New Developments in the Sciences*. George Braziller, New York. [ch.12]

Laszlo E. (1972b). *Introduction to Systems Philosophy: Toward a New Paradigm of Contemporary Thought*. Harper & Row, New York. [ch.12]

Laszlo E. (1977). *Goals for Mankind: A Report to the Club of Rome on the New Horizons of Global Community*. The Research Foundation of the State University, New York, USA. [ch.12]

Laszlo E. (1995). *Personal communication*. [ch.12]

Laszlo E. (2001). *Macroshift: Navigating the Transformation to a Sustainable World*. Berrett-Koehler Publishing, Inc., San Francisco, CA, USA. [ch.12]

Laszlo E. (2003). *You Can Change the World: The Global Citizen's Handbook for Living on Planet Earth*. Select Books, New York. [ch.12]

Latour B. (1987). *Science in Action - How to Follow Scientists and Engineers through Society*. Harvard University Press, Cambridge, Massachusetts, USA. [ch.8]

Latour B. (1999). On recalling ANT. *Actor Network Theory and After*, J. Law, J. Hassard, eds., Blackwell Publishers/The Sociological Review, Oxford, UK, pp.15-25. [ch.1]

Le Moigne J.L. (1989). Systems profile: first, joining. *Systems Research*, 6:331-343. [ch.3]

Leopold A. (1949). *A Sand County Almanac: And Sketches Here and There*. Oxford University Press. [ch.16]

Levesque H.J. and Brachman R.J. (1985). A fundamental tradeoff in knowledge representation and reasoning. *Readings in Knowledge Representation*, R.J. Brachman, H.J. Levesque, eds., Morgan Kaufmann, Los Altos, California, pp. 41-70. [ch.1,2]

Lewis P. (1994). *Information Systems Development, Systems Thinking in the Field of Information Systems*. Pitman Publishing, London. [ch.9]

Lewis P.J. (1992). Rich picture building in the soft systems methodology. *European Journal of Information Systems*, 1:351-360. [ch.5]

Lombardi P.L. (2001). Responsibilities towards the coming generations: forming a new creed. *Urban Design Studies*, 7:89-102. [ch.2,4,11]

Luhmann N. (1970). Soziologie als Theorie sozialer Systeme. *Soziologische Aufklärung*, **Band 1**:113-136. Westdeutscher Verlag, Opladen, Germany. [ch.12]

Luhmann N. (1984). *Soziale Systeme: Grundriss einer allgemeinen Theorie*. Frankfurt am Main: Suhrkamp. English version: *Social Systems*, tr. Bednarz J. Jr., with Baecker D. Stanford University Press, Stanford, CA, USA, 1995. [ch.16]

Lyon D. (1988). *The Information Society: Issues and Illusions*. Polity Press, Cambridge, UK. [ch.6]

Lyon D. (1997). The Internet: beyond ethics? *Science and Christian Belief*. **9** (1): 35-45. [ch.8]

Lyytinen K. (1986). *Information systems development as social action: framework and critical implications*. Ph.D. thesis, Department of Computer Science, University of Jyvaskyla, Finland. [ch.10]

Lyytinen K. (1988). Expectation failure concept and systems analysts' view of information system failures: results of an exploratory study. *Information & Management*, **14**:45-57. [ch.11]

Lyytinen K. (1992). Information systems and critical theory. *Critical Management Studies*, M. Alvesson, H. Willmott, eds., Sage, London, pp. 159-180. [ch.10]

Lyytinen K. and Hirschheim R. (1987). Information systems failures - a survey and classification of the empirical literature. *Oxford Surveys in Information Technology*, 4:257-309. [ch.1,11]

Löwgren J. and Stolterman E. (1998). *Design av Informationsteknik*. Studentlitteratur, Lund, Sweden. [ch.3]

Mander J. (1996). Technologies of globalisation. *The Case Against the Global Economy - And for a Turn Towards the Local*, J. Mander, E. Goldsmith, eds., Sierra Book Clubs, San Franscisco, pp. 344-360. [ch.7]

Marx K. (1972 [1883]). *Das Kapital; Kritik der politischen Ökonomie, I.* Marx-Engels Werke, Berlin, Band 23. [ch.6]

Mason R.O. and Mitroff I.I., (1981). *Challenging Strategic Planning Assumptions*. Wiley, New York. [ch.4,13]

Maturana H.R. and Varela F.J. (1980). *Autopoiesis and Cognition: The Realization of the Living*. D. Reidel, Boston, USA. [ch.16]

May L. and Hoffman S. (1991). *Collective Responsibility: Five Decades of Debate in Theoretical and Applied Ethics*. Rowman & Littlefield, Savage, MD, USA. [ch.12]

McLuhan M. (1964). *Understanding Media: The Extensions of Man*. MIT Press, MA, USA. [ch.8]

Midgley G. (1994). Ecology and the poverty of humanism: a critical systems perspective. *Systems Research*, 11:67-76. [ch.15]

Midgley G. (1997). Mixing methods: developing Systemic Intervention. *Multimethodology: The Theory and Practice of Combining Management Science Methodologies*, J. Mingers, A. Gill, eds., John Wiley & Sons Ltd., Chichester, UK, pp. 249-290. [ch.3]

Midgley G. (2000). *Systemic Intervention: Philosophy, Methodology, and Practice.* Kluwer/Plenum, New York. [ch.15, Preface]

Midgley G. (2002). Renewing the critique of the theory of knowledge-constitutive interests: a reply to Reynolds. *Journal of the Operational Research Society*, 53:1165-1169. [ch.15]

Miller J.G. (1978). *Living Systems*. McGraw-Hill, New York. [ch.15]

Mingers J. (1980). Towards an appropriate social theory for applied systems thinking: critical theory and soft systems methodology. *Journal of Applied Systems Analysis*, 7:41-49. [ch.3]

Mingers J. (1984). Subjectivism and soft systems methodology - a critique. *Journal of Applied Systems Analysis*, 11:85-103. [ch.3]

Mingers J. (1990). The what/How distinction and conceptual models: a reappraisal. *Journal of Applied Systems Analysis*, 17:21-28. [ch.3]

Mingers J. (1992). Questions and suggestions in using soft systems methodology. *Systemist*, 14:54-61. [ch.3]

Mingers J. (1995). Using soft systems methodology in the design of information systems. *Information Systems Provision: The Contribution of Soft Systems Methodology*, F. Stowell, ed., Mc Graw-Hill, London / New York, pp. 18-51. [ch.1]

Mingers J. and Gill A. (1997). *Multimethodology: The Theory and Practice of Combining Management Science Methodologies*. John Wiley & Sons, New York / London. [ch.14]

Mingers J. and Taylor, S. (1992). The use of soft systems methodology in practice. *Journal of the Operational Research Society*, 43:321-332. [ch.5]

Minsky M. (1981). A framework for representing knowledge. *Mind Design*, J. Haugeland, ed., MIT Press, MA, USA, pp. 95-128. [ch.2]

Mirijamdotter A. (1998). *A Multi-Modal Systems Extension to Soft Systems Methodology.* Doctoral Thesis, Luleå University of Technology, Luleå, Sweden. http://epubl.luth.se/1402-1544/1998/06/LTU-DT-9806-SE.pdf [ch.3,5,10]

Mirijamdotter A. and Somerville M.M. (2004). Systems thinking in the workplace: implications for organizational leadership. *Proceedings of the Third International Conference on Systems Thinking in Management*. Philadelphia, USA, May 19-21, 2004. http://www.acasa.upenn.edu/icstm04/ [ch.5]

Miser H.J. and Quade E.S. (1985). *Handbook of Systems Analysis: Craft Issues and Procedural Choices*. Wiley, New York. [ch.13]

Mitcham C. (1994a). *Thinking through Technology: The Path between Engineering and Philosophy*. The University of Chicago Press, Chicago, USA. [ch.1,14, Preface]

Mitcham C. (1994b). Engineering design research and social responsibility. *Ethics of Scientific Research*, K. Shrader-Frechette, Rowman and Littlefield, Lanham, MD, USA, pp. 153-168 and 221-223. [ch.16]

Mitev N.N. (1996). More than a failure? the computerized reservation systems at French Railways. *Information Technology and People*, 9(4):8-19. [ch.11]

Mitev N.N. (2001). The social construction of IS failure: symmetry, the sociology of translation and politics. *(Re-)Defining Critical Research in Information Systems*, A. Adam, D. Howcroft, H. Richardson, B. Robinson, eds., University of Salford, Salford, UK, pp.17-34. [ch.11]

Morgan G. (1986). *Images of Organization*. Sage, London. [ch.3]

Mumford L. (1967). *The Myth of the Machine, I. Technics and Human Development.* Harcourt, Brace and World, Inc., New York. [ch.6]

Mumford L. (1971). *The Myth of the Machine, II. The Pentagon of Power.* Harcourt, Brace and World, Inc., New York. [ch.6]

Naughton J. (1979). Functionalism and systems research: a comment. *Journal of Applied Systems Analysis*, 6:69-73. [ch.3]

Neve T.O. (2003). Right questions to capture knowledge. *Electronic Journal of Knowledge Management*, 1(1):47-54 [ch.4]

Newbigin L. (1986). *Foolishness to the Greeks: The Gospel and Western Culture*. Eerdmans, Grand Rapids, MI, USA. [ch.12]

Ngwenyama O. (1987). *Fundamental Issues of Knowledge Acquisition: Toward a Human Action Perspective of Knowledge Acquisition*. Doctoral dissertation, Watson School of Engineering, State University of New York, Binghamton, NY, USA. [ch.10]

Noble D.F. (1999). *The Religion of Technology: The Divinity of Man and the Spirit of Invention*. Penguin Books. [Preface]

Nosek J.T. and McNeese M.D. (1997). Issues for knowledge management from experiences in supporting group elicitation & creation in ill-defined, emerging situations. *Proceedings of AAAI Spring Symposium on Artificial Intelligence in Knowledge Management*, March 24-26, 1997. Stanford University, USA. [ch.4]

Oliga J. (1996). *Power, Ideology, and Control*. Plenum, New York. [ch.15]

Pacey A. (2000). *The Culture of Technology*. MIT Press, Cambridge, MA, USA. [ch.1,6]

Parsons T. (1951). *The Social System*. Free Press, Glencoe, IL, USA. [ch.16]

Pascal B. (2000). *Pensées*. Librairie Générale Française, Paris. [ch.8]

Pellegrino E.D. and Thomasma D.C. (1981). *A Philosophical Basis of Medical Practice: Toward a Philosophy and Ethic of the Healing Professions*. Oxford University Press. [ch.12]

Pellegrino E.D. and Thomasma D.C. (1988). *For the Patient's Good: The Restoration of Beneficence in Health Care*. Oxford University Press. [ch.12]

Perin C. (2005). *Shouldering Risks: The Culture of Control in the Nuclear Power Industry*. Princeton University Press, Princeton, NJ, USA. [ch.16]

Pirsig R.M. (1981). *Zen and the art of motorcycle maintenance. An inquiry into values*. Bantam Books, New York. [ch.10]

Polanyi K. (1944). *The Great Transformation*. Farrar and Rinehart, New York. [ch.16]

Polanyi K. (1957). Aristotle discovers the economy. *Trade and Market in the Early Empires: Economies in History and Theory*, K. Polanyi, C.M. Arensberg, H.W. Pearson, eds., Free Press, Glencoe, IL, USA, pp. 64-94. [ch.16]

Polanyi M. (1967). *The Tacit Dimension*. Routledge and Kegan Paul, London. [ch.2,4]

Popper K.R. (1959 [1935]). *The Logic of Scientific Discovery*. Originally published as *Logik de Forschung*. Harper, New York. [ch.15]

Rapp F. (1974). *Contributions to a Philosophy of Technology: Studies in the Stucture of Thinking in the Technological Sciences*. D. Reidel, Dordrecht, Netherlands / Boston, USA. [ch.14]

Ray L.J. (1993). *Rethinking Critical Theory; Emancipation in the Age of Global Social Movements*. Sage, london. [ch.4]

Reid J.I., Gray D.I., Kelly T.C., Kemp E.A. (1999). An application of SSM in the on-Farm labour situation in the New Zealand dairy industry. *Systems Research and Behavioral Science*, 16:341-350. [ch.3]

Ropohl G. (1979). *Eine Systemtheorie der Technik: Zur Grundlegung der Allgemeinen Technologie*. Carl Hanser Verlag, München/Wien, Germany. [ch.12]

Rose J. (1997). Soft systems methodology as a social science research tool. *Systems Research*, 14:249-258. [ch.3]

Roy O. (2002). *L'islam Mondialisé*. Éditions du Seuil, Paris. [ch.7]

Sachs W. (1990). *Die Liebe zum Automobil - Ein Rückblick in die Geschichte unserer Wünsche.* Rowohlt-Taschenbuch-Verlag, Reinbek, Germany. [ch.7]

Sachs W. (1999). *Planet Dialectics: Explorations in Environment and Development.* Zed Books, London, New York. [ch.6]

Schregenberger J. (1982). The development of Lancaster Soft Systems Methodology: a review and some personal remarks from a sympathetic critic. *Journal of Applied Systems Analysis,* **9**:87-98. [ch.3]

Schuurman E. (1980). *Technology And Future - A Philosophical Challenge.* Wedge Publishing Foundation, Toronto. [ch.8]

Seashore, S.E. (1976). The design of action research. *Experimenting with Organizational Life: The Action Research Approach,* A. Clarke, ed., Plenum, New York, pp.103-117. [ch.3]

Senge P.M. (1990). *The Fifth Discipline: The Art and Practice of the Learning Organization.* Doubleday, New York / Century Business, London. [ch.5,16]

Shapley D. (1993). *Promise and Power: The Life and Times of Robert McNamara.* Little Brown, Boston, USA. [ch.16]

Simon H.A. (1996). *The Sciences of the Artificial,* 3rd ed. The MIT Press, Cambridge, MA, USA. [ch.6,14]

Smyth D.S. and Checkland P.B. (1976). Using a systems approach: the structure of root definitions. *Journal of Applied Systems Analysis,* **5**:75-83. [ch.3,5]

Somerville M.M., Huston, M., Mirijamdotter, A. (2005). Building on what we know: staff development in the digital age. *The Electronic Library,* **23**(4). [ch.5]

Somerville M.M. and Vazques, F. (2004). Constructivist workplace learning: an idealized design project. *Proceedings of the 3rd International Lifelong Learning Conference,* P.A. Danaher, C. Macpherson, F. Nouwens, D. Orr, eds., Central Queensland University, Rockhampton, Queensland, Australia, pp. 300-305. [ch.5]

Spinosa C., Flores F., Dreyfus H.L. (1997). *Disclosing New Worlds: Entrepreneurship, Democratic Action, and the Cultivation of Solidarity.* The MIT Press, Cambridge MA, USA. [ch.14]

Stolterman E. (1991). *Designarbetets Dolda Rationalitet.* Department of Information Processing, Umeå University, Umeå, Sweden. [ch.3]

Strijbos S. (1988). *Het Technische Wereldbeeld: Een wijsgerig onderzoek van het systeemdenken* [The Technical Worldview: A Philosophical Investigation of Systems Thinking]. Buijten and Schipperheijn, Amsterdam. [ch.14,16]

Strijbos S. (1994). The individual and the collective in health care: a problem of systems ethics. *Systems Research,* **11**:67-76. [ch.12]

Strijbos S. (1995). How can systems thinking help us in bridging the gap between science and wisdom? *Systems Practice,* **8**(4):361-376. [ch.10,13]

Strijbos S. (1996). Ethics for an age of social transformation, Part I: framework for an interpretation, Part II: the idea of a systems ethics. *World Futures,* **46**:133-143,145-155. [ch.14]

Strijbos S. (1997). The paradox of uniformity and plurality in technological society. *Technology In Society,* **19**(2):177-95. [ch.6,14]

Strijbos S. (1998). Ethics and the systemic character of technology. *Techne: Electronic Journal of the Society for Philosophy and Technology,* **3**(4):19-35. [ch.14]

Strijbos S. (2000). Systems methodologies for managing our technological society: towards a 'Disclosive Systems Thinking", *Journal of Applied Systems Studies,* **1**(2):159-181. [ch.10]

Strijbos S. (2001). Global citizenship and the real world of technology. *Technology In Society*, **23**:525-533. [ch.6]

Strijbos S. (2004). Towards a new interdisciplinarity: a discussion paper. *Towards a New Interdisciplinarity - Proceedings of the 9th Annual Working Conference of CPTS*, R. Nijhoff, S. Strijbos, B. Bergvall-Kåreborn, A. Mirijamdotter, eds., Centre for Philosophy, Technology and Social Systems, Institute for Cultural Ethics, Amersfoort, Netherlands, pp. 133-138. [ch.1]

Susman G. (1983). Action research: a sociotechnical systems perspective. *Beyond Method: Strategies for Social Research*, G. Morgan, ed., Sage, London, pp. 95-113. [ch.3]

Tarnas R. (1991). *The Passion of the Western Mind*. Pimlico, Random House, New York. [ch.11]

Taylor Ch. (1991). *The Ethics of Authenticity*. Harvard University Press, Cambridge, MA, USA. [ch.7]

Taylor Ch. (1995). Comparison, history, truth. *Philosophical Arguments*, Taylor Ch., ed., Harvard University Press, Cambridge, MA, USA, pp. 146-164. [ch.8]

The Holy Bible (1989). *The New Revised Standard Version*. Oxford University Press. [ch.12]

Toffler A. (1980). *The Third Wave*. Collins, London. [ch.12]

Troncale L.R. (1985). The future of general systems research: Obstacles, potentials, case studies. *Systems Research*, **2**:43-84. [ch.15]

Tsouvalis C. and Checkland, P. (1996). Reflecting on SSM: the dividing line between 'real world' and systems thinking world. *Systems Research*, **13**:35-45. [ch.3]

Turkle S. (1995). *Life on the Screen - Identity in the Age of the Internet*. Simon & Schuster. New York. [ch.7,8]

Ulrich W. (1981) A critique of pure cybernetic reason: The Chilean experience with cybernetics. *Journal of Applied Systems Analysis*, **8**:33. [ch.13]

Ulrich W. (1983). *Critical Heuristic of Social Planning: A New Approach to Practical Philosophy*. Haupt, Bern, Switzerland. [ch.9,10,13,15]

Ulrich W. (1987). Critical heuristics of social system design. *European Journal of Operation Research*, **31**:276. [ch.10,13]

Ulrich W. (1988). Systems thinking, systems practice, and practical philosophy: a program of research. *Systems Practice*, 1(2):137-163. [ch.9]

Ulrich W. (1991). Systems thinking, systems practice, and practical philosophy: a program of research. *Critical Systems Thinking: Directed Readings*, Flood R.L., Jackson M.C., eds., John Wiley & Sons, Chicester, UK, chapter 12. [ch.14]

Ulrich W. (2001a). A philosophical staircase for information systems definition, design, and development: a discursive approach to reflective practice in ISD (part 1). *Journal of Information Technology Theory and Application*, **3**:55-84. [ch.15]

Ulrich W. (2001b). Critically systemic discourse: A discursive approach to reflective practice in ISD (part 2). *Journal of Information Technology Theory and Application*, **3**:85-106. [ch.15]

Ulrich W. (2003). Beyond methodology choice: critical systems thinking as critically systemic discourse. *Journal of the Operationsal Research Society*, **54**:325-342. [ch.13]

Van der Kooy T.P. (1953). *Op het Grensgebied van Economie en Religie*. Gebr. Zomer en Keunings, Wageningen. [ch.14]

Van der Stoep J. (1998). Internet: a game without rules?. *Proceedings of the 42th Annual Conference of the International Society of the Systems Sciences. Atlanta, Georgia*. [ch.8, Preface]

Van der Stoep J., Kee, B. (1997). Hypermobility as a challenge for systems thinking and government policy. *Systems Research and Behavioural Science*, 14:399-408. [ch.7, Preface]

Van Dijk J.A.G.M. (1997). *De Netwerkmaatschappij - Sociale Aspecten van Nieuwe Media*. Bohn Stafleu Van Loghum, Houten, Netherlands. [ch.8]

Van Riessen H. (1949). *Filosofie en Techniek* [Philosophy and Technology]. Kok, Kampen, Netherlands. [ch.6]

Van Riessen H. (1952,1957). *Society of the Future*. Presbyterian and Reformed Publishing Company, Philadelphia. [ch.6]

Van Riessen H. (1970). *Wijsbegeerte* [Philosophy]. Kok, Kampen, Netherlands. [ch.8]

Van Riessen H. (1979a [1961]) The structure of technology. *Research in Philosophy & Technology*, 2:296-313. [ch.6,14]

Van Riessen H. (1979b [1951]) Technology and culture. *Research in Philosophy & Technology*, 2:313-328. [ch.6]

Vickers G. (1959). *The Undirected Society*. University of Toronto Press, Toronto. [ch.3]

Vickers G. (1972). *Freedom in a Rocking Boat: Changing Values in an Unstable Society*. Penguin, Harmondsworth, UK. [ch.15]

Virilio P. (1998). *La Bombe Informatique*. Galilée, Paris. [ch.7,8]

Visser L., Vlug A.E., Van der Lei J., Stricker B.H.Ch. (1996). Cough due to ace-inhibitors: a case-control study using automated general practice data. *Pharmacoepidemiology and Prescription*, 49:439-444. [ch.9]

Von Bertalanffy L. (1956). General system theory. *General Systems*, 1:1-10. [ch.15]

Von Bertalanffy L. (1968). *General System Theory: Foundations, Development, Applications*. Penguin Books, UK / G. Braziller, New York. [ch.6,10,16]

Von Bertalanffy L. (1975). The history and development of General System Theory. *Perspectives on General System Theory: Scientific-Philosophical Studies*. George Braziller, New York, pp. 149-169. [ch.14]

Von Bülow I. (1989). The bounding of a problem situation and the concept of a system's boundary in soft systems methodology. *Journal of Applied Systems Analysis*, 16:35-41. [ch.3]

Walsh B.J. and Middleton J.R. (1984). *The Transforming Vision: Shaping a Christian World View*. IVP, Illinois, USA. [ch.1]

Walzer M. (1983). *Spheres of Justice - A Defense of Pluralism and Equality*. Basic Books, New York. [ch.8]

Wand Y. and Weber R. (1995). On the deep structure of information systems. *Information Systems Journal*, 5(3):203-224. [ch.2]

Warfield J.N., Cardenas A.R. (1994). A handbook of interactive management. Iowa State University Press, Ames, IA, USA. [ch.13]

Warmington A. (1980). Action research: its method and its implications. *Journal of Applied Systems Analysis*, 7:23-39. [ch.3]

Weber M. (1949). *The Methodology of the Social Sciences*. Glencoe, Illinois, USA. [ch.5]

Weick K.E. (1995). *Sensemaking in Organizations*. Sage, London. [ch.5]

Whyte G. and Bytheway A. (1996). Factors affecting information systems' success. *International Journal of Service Industry Management*, 7:74-93. [ch.11]

Wiener N. (1948). *Cybernetics: Or, Control and Communication in the Animal and the Machine*. Wiley, New York. [ch.16]

Wiener N. (1954). *The Human Use of Human Beings: Cybernetics and Society*, 2nd revised ed. Doubleday, Garden City, NY, USA. [ch.6]

Willmott H. (1995). The odd couple: reengineering businesses; managing human relations. *New Technology, Work and Employment*, **10**:89-98. [ch.15]

Wilson B. (1990,1992). *Systems: Concepts, Methodologies and Applications*, (2nd ed., reprinted 1991. John Wiley and Sons, Chichester, UK. [ch.3,5]

Wilson B. (2001). *Soft Systems Methodology: Conceptual Model Building and its Contribution*. John Wiley and Sons, Chichester, UK. [ch.5]

Wilson E.O. (1998). *Consilience: The Unity of Knowledge*. Random House, New York. [ch.16]

Winfield M.J. (2000). *Multi-Aspectual Knowledge Elicitation*. PhD Thesis, University of Salford, UK. [ch.2,4,11, Preface]

Winfield M.J., Basden A., Cresswell I. (1995). A revised multi-Modal approach to information systems design. *Proceedings of the Thirty Ninth Annual Meeting of the International Society for Systems Thinking*, July 24-28, 1995. Free University, Amsterdam. [ch.4,10, Preface]

Winfield, M.J., Basden A. and Cresswell, I. (1996). Knowledge elicitation using a multi-modal approach. *World Futures*, **47**:93-101. [ch.5, Preface]

Winner L. (1990). Engineering ethics and political imagination. *Broad and Narrow Interpretations of Philosophy of Technology*, P.T. Durbin, ed., Kluwer Academic Publishers, Dordrecht, Netherlands / Boston, USA, pp. 53-64. [ch.7]

Winter M.C., Brown D.H., Checkland, P.B. (1995). A role for soft systems methodology in information systems development. *European Journal of Information Systems*, **4**:130-142. [ch.3]

Wolters A. (1993). *Creation Regained: Biblical Basics for a Reformational Worldview*. W.B. Eerdmans Publishing Company, Grand Rapids, MI, USA. [ch.1]

Wolters A.M. (1986). *Creation Regained: A Transforming View of the world*. Inter-Varsity Press, Leicester, UK. [ch.14]

Yates-Mercer P. and Bawden D. (2002). Managing the paradox: the valuation of knowledge and knowledge management. *Journal of Information Science*, **28**(1):19-29. [ch.4]

Young I.M. (2000). *Inclusion and Democracy*. Oxford University Press. [ch.8]

INDEX

Note: Names of authors may be found in the References section, which indicates the chapters in which they occur. Names of persons are only included in this index where they are treated as actors, such as historical figures, rather than as referenced authors.